A First Course in Dimensional Analysis

A First Course in Dimensional Analysis

Simplifying Complex Phenomena Using Physical Insight

Juan G. Santiago

The MIT Press
Cambridge, Massachusetts
London, England

This book was set in Times New Roman by Westchester Publishing Services. Printed and bound in the United States of America.

Library of Congress Cataloging-in-Publication Data

Names: Santiago, Juan G., author.
Title: A first course in dimensional analysis : simplifying complex phenomena using physical insight / Juan Santiago.
Description: Cambridge, MA : The MIT Press, [2019] | Includes bibliographical references and index.
Identifiers: LCCN 2018060953 | ISBN 9780262537711 (pbk. : alk. paper)
Subjects: LCSH: Dimensional analysis—Textbooks. | Mathematical analysis—Textbooks.
Classification: LCC QC20.7.D55 S26 2019 | DDC 530.8—dc23 LC record available at https://lccn.loc.gov/2018060953

10   9   8   7   6   5   4   3   2   1

To Philip, my father, who loved to build, and to Philip, my son, who loves to invent

Contents

Preface   xi

1   **A Brief History of Dimensional Analysis   1**

2   **Preliminary Concepts   5**
    2.1   Physical Quantities   5
    2.2   Dimensions of Terms   6
    2.3   Unit Conversion Factors   8
    2.4   Notation and Manipulation of Unknown Functions   9
        2.4.1   Notations for Unknown Functions   9
        2.4.2   Useful Concepts Involving Unknown Functions   11
    2.5   Principle of Dimensional Homogeneity   11
        2.5.1   Check for Errors in a Derivation   12
        2.5.2   Guessing a Function from the Dimensions of Its Variables: How Much of This Martini Have I Had?   13
    2.6   Approximations and Asymptotic Limits   14
    2.7   Summary   15
        Problems   16

3   **Some Basic Physical Principles   19**
    3.1   Newton's Second Law   19
    3.2   Conservation of Energy in Dynamics   20
    3.3   Basic Concepts of Fluid Flow   21
    3.4   Summary   24
        Problems   25

4   **Dimensional Analysis: Motivation and Introduction   27**
    4.1   Geometric Similarity   28
    4.2   The Drawbacks of Brute Force Experimentation: We Need a Bigger Submarine   29
    4.3   Comments on the Submarine Example   31
    4.4   Drag Coefficient as a Tool to Reduce Complexity   32
        4.4.1   Effect of Streamlining   32
        4.4.2   Drag on a Smooth Sphere: A First Example of Data Collapse   33
        4.4.3   Drag and Terminal Velocity of a Skydiver   35

4.5   Summary   38
        Problems   39

5   Dimensional Analysis Techniques   41
5.1   Rules of Thumb for Initial Hypothesized Function: What to Include or Exclude?   41
        5.1.1   Rule D1: Formulate in Terms of Known Algebraic Combinations   42
        5.1.2   Rule D2: Exclude a Variable Expressible as a Function of the Others   43
        5.1.3   Rule D3: Keep Dimensional Constants but Absorb Nondimensional Constants   43
        5.1.4   Rule D4: Exclude Any Variable That Involves a Unique Dimension (No Blood from a Rock)   44
        5.1.5   Rule D5: If Invoked, Consistently Leverage Geometric Similarity   44
5.2   Ipsen's Method: A Step-by-Step Process   45
5.3   Submarine Example Revisited   47
5.4   An Inelegant Application of Ipsen's Method   49
5.5   Time for a Stone to Drop: Experimental Closure and Collapse of Data   50
        5.5.1   Supplementing Dimensional Analysis with Experiments   52
        5.5.2   A Note on Guiding the Nondimensional Parameters   54
5.6   From Stones and Earth to Planets and Stars   55
5.7   Ignoring the Rule That We Exclude Determined Parameters: Including Both G and g 58
5.8   Period of a Pendulum for Any Angle: Experimental Closure and Collapse of Data   59
5.9   Summary   63
        Problems   64

6   Combining Dimensional Analysis with Physical Intuition and Experimental Observations   67
6.1   Rules of Thumb for Manipulating Functions of Nondimensional Variables   67
        6.1.1   Rule ND1: Reorganize Expressions of Nondimensional Variables   68
        6.1.2   Rule ND2: Isolate, Then Evaluate, Known Dependence   69
        6.1.3   Rule ND3: Isolate a Variable with an Unknown but Weak Dependence   71
        6.1.4   Rule ND4: Replace a Nested Function with Its Independent Nondimensional Parameters   72
        6.1.5   Rule ND5: Consider Absorbing an Approximately Constant Nondimensional Variable into Function   73
        6.1.6   Rule ND6: Be Careful Eliminating Variables—Even If They Are Small   73
6.2   Spine Patterns of Liquid Drop Impacts and Blood Drop Patterns in Forensics   74
6.3   Atomic Explosions: An Example Confirmation of a Power Law   78
        6.3.1   History of Atomic Explosion Analysis by G. I. Taylor   78
        6.3.2   Atomic Explosion Analysis Using Dimensional Analysis   80
6.4   World-class Weightlifters: Lifter-to-Lifter Comparisons and Lift Data Collapse   82
6.5   Running (from) Dinosaurs: An Analysis That Does Not Assume Geometric Similarity   84
6.6   Summary   90
        Problems   91

7   Two Examples Combining Biomechanics and Fluid Mechanics   93
7.1   Olympic Rowers and a Word of Caution on the Experimental Validation of Scaling Laws   93

7.2   A Derivation for the Great Flight Diagram: Approximate Data Collapse   99
        7.2.1   The Great Flight Diagram   103
7.3   Justification for Excluding Flier Power from the Formulation   105
7.4   Including Viscosity in the Flight Analysis   106
7.5   Summary   108
        Problems   109

8   Buckingham Pi Theorem: An Alternate Method   111
8.1   Buckingham Pi Theorem   111
8.2   Pressure Drop in a Pipe Explored Using the Buckingham Pi Theorem   112
8.3   Use of a Large Number of Scaled Pressure Drop Measurements to Close the
        Problem   116
8.4   Buckingham Pi Theorem versus Ipsen's Method   119
8.5   Summary   119
        Problems   120

9   Leveraging of Model Data to Build and Understand Prototypes   121
9.1   Geometric, Kinematic, and Dynamic Similarity   121
9.2   Experimental Design and Interpretation: To Match All Variables, Match All but One   122
9.3   Examples of Model and Prototype Studies   123
        9.3.1   Submarine Prototype Scaling   124
        9.3.2   Boat Drag and How Scale Model Studies Are Not Always Possible   125
9.4   Summary   128
        Problems   129

**10   Small Changes in Geometry Can Have Significant Effects: The Effect of Roughness
       on Drag   131**
10.1   Roughness and the Drag of Spheres   131
10.2   Pressure Drop in Smooth versus Rough Pipes   135
10.3   Estimating Drag from Different Shapes   135
10.4   Example of Incorrect Dimensional Analysis: Neglecting a Parameter   136
10.5   Summary   138
        Problems   139

11   The Riabouchinsky–Rayleigh Paradox and the Rule of Relevance   141
11.1   The Riabouchinsky–Rayleigh Paradox   141
11.2   Resolution of the Apparent Paradox   143
11.3   The Rule of Relevance   144
11.4   Summary of Rules of Thumb for Combining Physical Insight with
        Dimensional Analysis   145

12   Common Dimensionless Groups   147
12.1   Mach Number   147
12.2   Euler Number   149
12.3   Table of Dimensionless Groups   149
12.4   Summary   150
        Problems   150

13   Scaling Using Approximate Equations   151
    13.1   Summary   153
           Problems   154

Closing Note   155
Appendix A: Properties of Common Fluids   157
References   159
Index   163

Definition I: The quantity of matter [mass] is the measure of the same arising from its density and bulk [volume] conjointly.

Definition II: The quantity of motion [momentum] is the measure of the same, arising from the velocity and quantity of matter conjointly.

—Isaac Newton, *Principia* (1687)

These are the first two formal definitions in the *Principia*, here as translated by Chandrasekhar (1995), and these perhaps mark the beginning of modern, precise understanding of units and dimensions. Lord Rayleigh called the method of dimensional analysis "the great principle of similitude" (Rayleigh, 1915) in a now-famed article by that title. On its surface, it is a look at the relationships between physical quantities by comparing their "units." In fact, it is a powerful and formalized method to analyze complex physical phenomena and reduce complexity. The method enables engineers and scientists to analyze problems for which they cannot pose, much less solve, governing equations. The method serves engineers and scientist in the formation of back-of-the-envelope estimates and the derivation of scaling laws for the design of machines and processes. It is a quest for the natural variables of a problem.

The principle has been applied successfully to the study of complex phenomena in many areas of science, including biology, engineering, physics, economics, and sports science. In this text, we will demonstrate its use in the following fields:

- *Astrophysics* in a derivation of Kepler's third law
- *Aerodynamics* in the design of airplanes and estimation of terminal velocity
- *Forensics* in the analysis of blood splatters
- *Dynamics* in the study of falling bodies and a finite-displacement pendulum
- *Sports science* in the analyses of world-class rowers and weightlifters
- *Hydraulics* in the analysis of fluid flow in pipes
- *Marine engineering* in the design of submarines and surface boats

- *Atomic energy* in the analysis of a fission bomb detonation
- *Biomechanics* in the analysis of animal gait and flight
- *Paleontology* in estimates of running and flying speeds of dinosaurs and a giant bird

We will use these examples to show how a combination of dimensional analysis and physical intuition can identify the essential characteristics of a physical process. My hope is that the book will inspire discovery and innovation.

1

A Brief History of Dimensional Analysis

This book is an introduction to a powerful tool known as dimensional analysis. The method is also termed "similarity analysis," and the concept referred to as "similitude." Dimensional analysis can be used to find patterns within and succinct explanations for physical phenomena. It is used to find functional relations that are independent of whether we have a large amount of each quantity or a small amount. It is a search for laws that are independent of measurement systems.

The history of dimensional analysis is linked with the history of science starting as early as the great English mathematician and physicist Sir Isaac Newton. Histories and developments of the subject are provided by Macagno (1971), Rott (1992), White (2016), Taylor et al. (2007), and Gibbings (2011), and only some material is summarized here.

In 1687, Isaac Newton described some rules of dimensional relations and analyses in the *Principia*. This included similarity in the trajectory shapes of bodies in motion, and indeed his theories of motion present a precise set of mathematical relations between the quantities of position, velocity, acceleration, momentum, and force. In around 1736, Swiss mathematician and physicist Leonhard Euler wrote extensive descriptions of units and dimensional reasoning, including the dimensions associated with Newton's second law, and so-called unit conversion factors—concepts essential to consistent descriptions of the physical world.

In 1822, Jean-Baptiste Joseph Fourier clarified greatly the concept of physical dimensions, introduced the concept of dimensional homogeneity in mathematical relations, pointed out that equations describing physical phenomena are independent of the choice of system of units, and applied similarity to describe and unify heat flow. Lord Rayleigh (b. 1842) provided perhaps the most influential breakthroughs in dimensional analysis in studies of hydrodynamic drag, flow stability, sound waves, and light scatter from particles. He also strongly advocated dimensional analysis. He demonstrated that the "principle of similitude" yields elegant scaling laws and demonstrated the method using examples from Newton's law of gravitation, drag on spheres, heat flow in thin fins, vibrating drops, and the vibration of Aeolian harp strings. Rayleigh argued in favor of the importance and appropriate use of the method (e.g., as an initial analysis of a problem) (Rayleigh, 1915):

I have often been impressed by the scanty attention paid even by original workers in physics to the great principle of similitude. It happens not infrequently that results in the form of "laws" are put forward as novelties on the basis of elaborate experiments, which might have been predicted a priori after a few minutes' consideration.

Rayleigh stopped short of teaching a formalized approach to dimensional analysis, but the French mathematician Aimé Vaschy in 1892, and independently the Russian mathematician Dimitri Riabouchinsky in 1911, each formulated and then reported methods that we now call the Pi theorem (named after and introduced to most of the community by Edgar Buckingham, 1914). Since this time, dozens of texts and papers on similarity and similitude have been written. Modern similitude has been strongly leveraged by researchers in and for the field of fluid mechanics, but it has been applied successfully in many other fields. White (2016) provides references to examples in biomechanics, astrophysics, economics, chemistry, clinical medicine, social sciences, and plant genetics, among others.

We shall not attempt to review here all of the history of similarity methods or applications, but instead offer a few select milestone contributions to dimensional analysis as summarized in table 1.1. Note the years listed in the table are approximate dates of contribution as suggested by Rott (1992), Gibbings (2011), and White (2016).

Table 1.1
Brief review of important contributions to dimensional analysis

Individual and approximate date of contribution	Summary of Contribution
Newton, 1687	Defines basic relationships between mass, momentum, and force. Discusses similarity in motion trajectories in the *Principia*.
Euler, 1765	Extensive discussions of physical units, unit conversion factors; in, e.g., *Theoria motus corporum solidorum seu rigidorum*.
Fourier, 1822	Dimensional homogeneity, physical quantities vs. units, similarity in heat flow in his *Analytical Theory of Heat*.
Stokes, 1856	Identifies Reynolds number as key parameter in drag effects on a pendulum.
Maxwell, 1871	Applies dimensional analysis to electromagnetic theory and theory of gases, introduces M, L, and T nomenclature.
Rayleigh, 1877	Application of similitude to wide range of problems, demonstrating wide generality of approach. Extensive applications to light scatter and fluid mechanics, and to correlating measurements of drag.
Reynolds, 1883	Turbulent transition experiments identify Reynolds number (named/highlighted by Sommerfeld, 1908 and later reemphasized and taught by Prandtl, 1910).
Vaschy, 1892, and Riabouchinsky, 1911	Independently develop formalized methods now known as Buckingham Pi theorem.
Buckingham, 1914	Publishes and teaches the Pi theorem.
Bridgman, 1922	Publishes classic book on dimensional analysis theory.
Ipsen, 1960	Ipsen's method, publishes a book describing a step-by-step approach to dimensional analysis.

The last entry of table 1.1 is a method for performing dimensional analysis due to David Carl Ipsen (1960), a professor at the University of California at Berkeley. Ipsen's method is a formal, step-by-step approach that we will primarily adopt in this book for the process of dimensional analysis.

In addition to Ipsen's method, we will discuss a set of rules of thumb that in this text I propose as a guide to the integration of physical intuition and experimental observations with dimensional analysis. Enabling the combination of dimensional analysis and physical intuition is a major goal of this book.

2

Preliminary Concepts

… for a stone projected is by the pressure of its own weight forced out of the rectilinear path, which by the projection alone it should have pursued, and made to describe a curve line in the air; and through that crooked way is at last brought down to the ground; and the greater the velocity is with which it is projected, the farther it goes before it falls to the earth.

—Isaac Newton, *Principia* (1687)

We first present some preliminary concepts as a brief background to our study of dimensional analysis. We discuss the underlying dimensions of physical problems, the relation of these dimensions to units of measurement, and some concepts useful in discussing unknown functions.

2.1 Physical Quantities

To determine a set of rules with which to describe physical phenomena, we first identify the quantities of interest. In aerodynamics, interesting quantities include velocity, wing area, weight, and power. In biomechanics, quantities of interest might include the height of an animal's center of mass, running speed, and the mechanical stress exerted by muscle tissue. In earthly dynamics, we may analyze the gravitational acceleration at Earth's surface, heights, velocities of moving bodies, and the lengths of pendulums. In astrophysics, we consider vast distances, orbit periods, and large masses. To record, communicate, and analyze information and compare observations at different times and in different places, we create scales of measure. For example, we agree on the value of 1 kilogram or the duration of 1 second.

However, if we are not careful, we can easily attribute too much importance to these scales of measure. The scales and amounts of each observation are incidental since underlying laws governing physical phenomena are independent of arbitrary unit definitions or specific amounts of each quantity. Nature does not care about these absolute measures, only the amounts relative to each other. The relations among quantities can be found from something very mundane that we learn in principle as children by looking at their dimensions. For example, velocity is a length per time, and so it must be relatable to relevant lengths and relevant times.

Following the work of French mathematician and physicist Jean-Baptiste Joseph Fourier (1822), we describe natural phenomena using relations of physical quantities and we must make "apples to apples" type comparisons. That is, only quantities measured along the same dimension can be compared directly and unambiguously. The comparison should be independent of the particular choice of units (e.g., meters or feet or furlongs). Scottish physicist James Clerk Maxwell (1871) suggested we quantify physical quantities in terms of universal dimensions with notation L, T, and M in dynamics. Here, as in many texts, we will adopt Maxwell's notation and describe length, time, mass, absolute temperature, and electric current as L, T, M, θ, and I, respectively. Table 2.1 provides examples of quantities characterized in terms of these dimensions in our list. For brevity, we neglect quantities involving luminosity (e.g., with units of candela).

L, T, M, θ, and I each represent a dimension associated with a physical quantity. They are in general independent and so can be described as orthogonal (e.g., one cannot be measured or obtained by knowing the others). These ideas can also be applied to groups of dimensions. The dimensions L^3 and MLT^{-1} quantify volume and momentum, respectively. We can plot the volume versus momentum of a ball on a two-dimensional axis of L^3 versus MLT^{-1} with, for example, corresponding units m^3 and $kg \cdot m \cdot s^{-1}$. In turn, these units are derived from understood meanings of m, kg, and s as particular measures of L, M, and T, respectively. In the case of table 1.1, we express units using the SI standard, the French Système internationale d'unités, or international system of units.

Note that the choice of L, T, M, θ, and I as primary dimensions is not unique. For example, we can alternately, and self-consistently, choose L, T, F, θ, and I as primary dimensions with F for force as a basic, primary dimension. Mass is then a dimension derived from L, T, and F (mass is FT^2L^{-1}).

2.2   Dimensions of Terms

An equation consists of an equal sign (=) and its terms. Terms are the items separated by addition, and these can be positive or negative. For example, for idealized motion in the x-direction under constant acceleration a_0:

$$x = x_0 + u_0 t + \frac{a_0 t^2}{2}, \tag{2.1}$$

where x_0 and u_0 are the initial position and velocity and t is time. This equation has four terms, three of them on the right-hand side, each with dimension of length. The dimensions of terms can be determined by dividing or multiplying the dimensions of each

Table 2.1
Example quantities, their primary dimensions, and corresponding (derived) SI units

Quantity	Primary dimension	SI units example
Mechanical quantities		
Acceleration	LT^{-2}	$m \cdot s^{-2}$
Angle (plane)	—	radian
Angle (solid)	—	steradian
Angular acceleration	T^{-2}	$rad \cdot s^{-2}$
Angular momentum	ML^2T^{-1}	$kg \cdot m^2 s^{-1}$
Area	L^2	m^2
Curvature	L^{-1}	m^{-1}
Surface tension	MT^{-2}	$kg \cdot s^{-2}$
Density	ML^{-3}	$kg \cdot m^{-3}$
Elastic modulus	$ML^{-1}T^{-2}$	$kg \cdot m^{-1}s^{-2}$
Energy (work)	ML^2T^{-2}	$kg \cdot m^2 \cdot s^{-2}$
Force	MLT^{-2}	$kg \cdot m \cdot s^{-2}$
Frequency	T^{-1}	s^{-1}
Kinematic viscosity	L^2T^{-1}	$m^2 \cdot s^{-1}$
Mass	M	kg
Momentum	MLT^{-1}	$kg \cdot m \cdot s^{-1}$
Power	ML^2T^{-3}	$kg \cdot m^2 \cdot s^{-3}$
Pressure	$ML^{-1}T^{-2}$	$kg \cdot m^{-1} \cdot s^{-2}$
Time	T	s
Velocity	LT^{-1}	$m \cdot s^{-1}$
Volume	L^3	m^3
Gravitational constant	$L^3M^{-1}T^{-2}$	$m^3 \cdot kg^{-1} \cdot s^{-2}$
Thermal properties		
Enthalpy	ML^2T^2	$kg \cdot m^2 \cdot s^2$
Entropy	$ML^2T^2\theta^{-1}$	$kg \cdot m^2 \cdot s^{-2} \cdot K^{-1}$
Heat capacity per unit mass	$L^2 T^{-2}\theta^{-1}$	$m^2 \cdot s^{-2} \cdot K^{-1}$
Internal energy	ML^2T^{-2}	$kg \cdot m^2 \cdot s^{-2}$
Latent heat of phase change	L^2T^{-2}	$m^2 \cdot s^{-2}$
Quantity of heat	ML^2T^{-2}	$kg \cdot m^2 \cdot s^{-2}$
Temperature	θ	K
Thermal conductivity	$MLT^{-3}\theta^{-1}$	$kg \cdot m \cdot s^{-3} \cdot K^{-1}$
Thermal diffusivity	L^2T^{-1}	$m^2 \cdot s^{-1}$
Heat transfer coefficient	$MT^3\theta^{-1}$	$kg \cdot s^3 \cdot K^{-1}$
Boltzmann constant	$ML^2T^{-2}\theta^{-1}$	$kg \cdot m^2 \cdot s^{-2} \cdot K^{-1}$
Electromagnetic quantities		
Electric current	I	$C \cdot s^{-1}$
Electric charge	IT	C
Electric field	$MLI^{-1}T^{-3}$	$kg \cdot m \cdot A^{-1} \cdot s^{-3}$
Electric resistance	$ML^3T^{-3}I^{-2}$	$kg \cdot m^3 \cdot s^{-3} \cdot A^{-2}$
Magnetic flux density	$MT^{-2}I^{-1}$	$kg \cdot s^{-2} \cdot A^{-1}$
Magnetic field	IL^{-1}	$A \cdot m^{-1}$
Magnetic susceptibility	—	—
Permittivity of free space	$M^{-1}L^{-3}I^2T^4$	$kg^{-1} \cdot m^{-3} \cdot A^2 \cdot s^4$
Permeability of free space	$MLT^{-2}I^{-2}$	$kg \cdot m \cdot s^{-2} \cdot A^{-2}$

Adapted and expanded from similar tables given by Yarin (2012) and Mills (1993).

variable in the term. The dimensions of integrals and derivatives follow from their definitions as shown in table 2.2. Integrals are sums of products of the integrand and differential, and so have the dimension of this product. Similarly, derivatives are limits of the ratio of differences and have dimensions of the ratio.

As one example, consider an integro-differential equation of a form common in an undergraduate fluid mechanics courses (e.g., derived by applying Newton's second law to a control volume with fluid moving through it):

$$\underbrace{\int_0^{A_0} p_G dA}_{term\ 1} + \underbrace{S}_{term\ 2} - \underbrace{m\frac{dv}{dt}}_{term\ 3} = \underbrace{\rho u^2 A_1}_{term\ 4} \tag{2.2}$$

Here, S, p_G, A, m, ρ, and t represent a force, pressure, area, mass, density, and time, respectively; u and v are fluid velocities.

2.3   Unit Conversion Factors

Introduced by Euler (circa 1765), unit conversion factors (UCFs) are a convenient method of changing from one set of units to another. Consider a solution to equation (2.1) for the case of $a_0 = 0$:

$$u_0 = \frac{x - x_0}{t}. \tag{2.3}$$

The three variables represent physical quantities with dimensions, so for $x - x_0 = 2.0$ ft and t of 2.0 s, we can write

Table 2.2
The dimensions of integrals and differentials

Quantity	Dimension
Indefinite Integral $\int f(x)dx$	Dimension of the product of $f(x)$ and x; e.g., if $f(x)$ has dimensions of velocity LT^{-1} and x dimensions of length L, then this integral will have dimensions L^2T^{-1}.
Definite Integral $\int_0^{x_0} f(x)dx$	Dimension of the product of $f(x)$ and x_0.
Derivative $df(x)/dx$	Dimension of the ratio $f(x)/x$; e.g., if $f(x)$ has dimension of velocity LT^{-1} and x dimension of time T, then this derivative will have dimension LT^{-2}.
Partial Derivative $\partial f(x)/\partial x$	Dimension of the ratio $f(x)/x$.

$$u_0 = \frac{x - x_0}{t} = 1.0 \text{ ft/s}.$$

The quantity on the right-hand side is not just "1," but 1 ft·s^{-1} (think of the units as determining the meaning of the 1.0). To change from English to SI units, we use the following UCF:

$$\text{UCF} = \frac{1 \text{ ft}}{0.3048 \text{ m}} = 1$$

The UCF is a quantity identically equal to the concept of one, which we call unity. Hence, its reciprocal is also unity:

$$\frac{0.3048 \text{ m}}{1 \text{ ft}} = 1.$$

Importantly, we can multiply one or both sides of the "=" sign by unity, hence:

$$u_0 = 1 \text{ f} \cdot \text{s}^{-1} \cdot \frac{0.3048 \text{ m}}{1 \text{ ft}} = 0.3 \text{ m} \cdot \text{s}^{-1}.$$

We can multiply, say, the right-hand side of any equation multiple times by unity and so employ multiple UCFs as follows:

$$V = \frac{5.0 \text{ ft}}{\text{min}} \cdot \underbrace{\frac{1 \text{ min}}{60 \text{ s}}}_{\text{UCF1}} \cdot \underbrace{\frac{12 \text{ in}}{1 \text{ ft}}}_{\text{UCF2}} \cdot \underbrace{\frac{25.4 \text{ mm}}{1 \text{ in}}}_{\text{UCF3}} = \frac{25.4 \text{ mm}}{\text{s}}.$$

2.4   Notation and Manipulation of Unknown Functions

Before introducing dimensional analysis, it is useful to gain a familiarity with the expression and description of mathematical functions. Here we will explore shorthand notations for functions that are (at least initially) unknown but that are known or believed to depend on some known set of variables.

2.4.1   Notations for Unknown Functions

In dimensional analysis, we typically deal with unknown relationships (unknown forms of an equation) among variables. For functions involving dimensional variables, we will write y as an unknown function of x as, for example,

$$y = f(x). \tag{2.4}$$

This notation stands for an infinite number of possible relations between x and y, where y is the dependent variable and x is the independent variable. We here use a lowercase

"f" for the function when it involves dimensional variables. Here f denotes "a function of" or "determined by." If there is a single value of y for each value of x, we say f is a single valued function. Colloquially, "Give me x, and I'll tell you y." For the function, $z = f(x, y)$, "Give me x and y, and I'll tell you z."

For functions of strictly nondimensional variables X and Y, we shall employ capital letters to denote a function, as in

$$Y = F(X).$$

Consider a more complex function between a velocity u, time t, distance x, and acceleration a of the form

$$u = f(t, x, a).$$

There are many possibilities. One is

$$u = \frac{x}{t} + \int_0^t a \, dt,$$

and another is

$$u = \frac{x_0}{t_0} \exp\left(-t\sqrt{\frac{a}{x}}\right).$$

where x_0 and t_0 are specific values of x and t (two constants). Each function is physically possible, both are represented by

$$u = f(t, x, a).$$

Special Case: A Function of a Constant   Consider the deterministic function

$$u = f(t).$$

If we can conclude from additional arguments that t is a constant, say, equal to t_0, then we can also conclude that the function is a constant

$$u = f(t_0) = u_0 = constant.$$

A deterministic function of a constant is constant. Note that the constant u_0 is nevertheless intimately and inextricably tied to the meaning of t_0 (it is now baked into the definition of f).

Throughout this text we shall consider single valued functions (for the dependent variable) of a series of independent variables. We shall invert the functions (e.g., to solve for a new variable) only if the physics of the problem suggest there is no ambiguity (such as from double or triple valued functions).

2.4.2   Useful Concepts Involving Unknown Functions

This section presents a list of useful examples and operations involving unknown functions.

Concept 1: A Nested Function   Consider some function

$$h = f_1(x, g).$$

If we can argue (through intuition or experiment) that the variable g is itself a function of x as in $g = f_2(x)$, then we can write simply

$$h = f_3(x).$$

We shall use this idea to eliminate variables from consideration in unknown functions.

Concept 2: A Range of Applicability   Important for physical problems, some functions bring with them a limited range of applicability, like baggage on a trip.

   Consider, for example, two functions of x and respective ranges of applicability:

$$f(g) = g^2 + 3 \text{ for } -\infty < g < \infty.$$

$$g(x) = x^{0.5} \text{ for } 0 \geq x < \infty.$$

Here, if we consider a nested function as in $h(x) = f(g(x))$, we are restricted to the intersection of the allowable domains for x and g, so

$$h(x) = x + 3 \text{ for } 0 \geq x < \infty.$$

   If the function of one variable is restricted to Tuesdays and a second function to morning time, then a function that depends on both of these functions is restricted to Tuesday mornings. This idea can help us avoid mistakes of applicability, such as deriving a negative mass.

2.5   Principle of Dimensional Homogeneity

As a preliminary exercise, consider two variables x and y and a three-term equation of the form

$$xy = x + y.$$

From a purely mathematical view, this equation looks reasonable. However, we should note that this function can hold only if both x and y are quantities without dimensions. Contrast this to the aforementioned equation (2.1):

$$x = x_0 + u_0 t + \frac{a_0 t^2}{2}.$$

This four-term equation is much more satisfying as a physical representation, and it should sit well in your stomach. Each of the terms has dimensions of length. In SI units, we write

$$\underset{m}{\underbrace{x}} = \underset{m}{\underbrace{x_0}} + \underset{m}{\underbrace{v_0 t}} + \underset{m}{\underbrace{\frac{a_0 t^2}{2}}}.$$

More generally, we write that each term has a primary dimension of length L such that

$$\underset{L}{\underbrace{x}} = \underset{L}{\underbrace{x_0}} + \underset{L}{\underbrace{v_0 t}} + \underset{L}{\underbrace{\frac{a_0 t^2}{2}}}.$$

These ideas bring us to a general principle applicable to all equations that aspire to represent the physical world.

The principle of dimensional homogeneity (PDH) was first described by Fourier (1822), and can be stated as follows (cf. White, 2016):

> If an equation truly expresses a proper relationship between variables in a physical process, then each additive term will have the same dimension.

Hence, both sides of the "=" sign must have the same dimension. The concept is simple, yet it is extremely powerful and general to all equations that represent the physical world. We can rephrase the PDH colloquially as "terms in equations must compare (i.e., add or equate) apples to apples."

Consider the units of equation (2.2), given that p is pressure, S is a force, m is mass, u is horizontal velocity, v vertical velocity, A is area, and t is time:

$$\underset{N}{\underbrace{\int_0^{A_0} p_G \, dA}} + \underset{N}{\underbrace{S}} - \underset{N}{\underbrace{m \frac{dv}{dt}}} = \underset{N}{\underbrace{\rho u^2 A_1}}.$$

You do not need to know about this equation's derivation to know that it must be dimensionally homogenous. Each term is "measured" in the dimension of force (see table 2.1).

2.5.1   Check for Errors in a Derivation

Among other uses, the PDH is extremely useful in spotting errors in derivation, such as an equation flashed on slide in a presentation. Consider equation (2.1) for the case where $x_0 = 0$ and $u_0 = 0$; we might mistakenly write something like

$$x = \frac{a_0^2 t}{2},$$

which has an obvious mistake. The dimensions don't work; the left-hand side has dimension L, and the right-hand side dimension is L^2T^{-3}, an apple and an orange.

Inside secret: Many university professors use this "trick" to quickly spot an error in a long derivation, often within a few seconds after it is projected on a slide. All that is needed is to understand (even vaguely) the variables, know their dimensions, and then errors in multipliers, exponents, and so on pop out. It's especially dramatic to wait for a seemingly impressive, complex final expression and quickly point out that there must be some error.

2.5.2 Guessing a Function from the Dimensions of Its Variables: How Much of This Martini Have I Had?

Sanjoy Mahajan (2008), in his nice monograph *Street-Fighting Mathematics*, provides some examples of using the PDH to estimate area and volume integrals of geometric shapes and other functions. Mahajan highlights the importance of preserving PDH and then considering limiting behaviors of the dependence.

Consider the relation between the free surface height and volume of the liquid in a martini glass, as depicted in figure 2.1. We approximate the liquid volume $\forall$ as a cone with radius r and liquid level, height h. Some guesses, including a couple of uninspired ones that nevertheless respect PDH, may include $\forall = c_1 r^4 / h$; $\forall = c_2 h^4 / r$; $\forall = c_3 h^3 + c_4 r^3$; $\forall = c_5 r h^2$; and $\forall = c_6 h r^2$. Here, the c_i's are arbitrary, dimensionless constants (pure numbers without units, such as π). Explore the limits of these relations. The first two expressions uncomfortably predict infinite volume for limits $h \to 0$ and $r \to 0$, respectively. The

Figure 2.1
A martini glass filled with liquid.

third is clearly incorrect for either $h \to 0$ or $r \to 0$. Our knowledge of triangles and the volumes of cylinders may help us reject $\forall = c_5 r h^2$, since the answer should be proportional to h. Consider stacking two cylinders of height $h/2$ atop one another—the total volume should double. The last supposition seems most likely, since $\forall = c_6 h r^2$ behaves well for either r or h tending to zero or infinity, and for both r or h tending to zero or infinity simultaneously. Also, the case $h = r$ looks fine. In fact, this guess is correct. To estimate the value of the constant c_6, consider the volume of a cylinder of height h: $\forall_{\text{cylinder}} = \pi h r^2$ (the area of a circle multiplied by the height). We know the cone will be smaller than this. With a little visualization, we might guess less than half this value, say one-third, as in

$$\forall_{\text{cone}} = \frac{\pi h r^2}{3}.$$

This is in fact the exact answer, but the exact answer does not really matter here; one-fourth or one-half of the cylinder's volume would be good working estimates. Further, to explore how the fraction of liquid consumed for some change in height, we need not know c_6. Martini glasses seem to favor interior angles of roughly 45°. So, for a martini glass we estimate that $h = r$ and might guess that

$$\forall_{\substack{\text{martini glass} \\ 45° \text{ angle}}} = \frac{\pi h r^2}{3} \overset{h=r}{=} \frac{\pi h^3}{3}.$$

Now, we can make estimates which follow from PDH and are completely independent of our guess of constant c_6. How much cocktail have we imbibed if we drink a discreet quantity halfway down from the initial height?

$$\text{fraction drunk} = \frac{\forall_{\text{initial}} - \forall_{\text{final}}}{\forall_{\text{initial}}} = \frac{h^3_{\text{initial}} - h^3_{\text{final}}}{h^3_{\text{initial}}} = 1 - \frac{h^3_{\text{final}}}{h^3_{\text{initial}}} = 1 - \left(\frac{1}{2}\right)^3 = 87.5\%.$$

Sipping the top half of the initial liquid cone's height, we drank 88% of the whole.

2.6   Approximations and Asymptotic Limits

In our martini example, we found it useful to determine the limiting behavior of a function as one or more of the independent variables tended to infinity or zero. We can think of this process as testing our function by asking of it interesting questions.

Consider a function's dependent variable as one or more of the independent variables tends to a limiting value. Consider first the case of function of a single variable. We express this as

$$y = f(x).$$

Using analysis, experiments, and intuition, we should consider the value or values of y as x tends to some critical value such as zero or infinity. There four possibilities as follows (see also the discussion in Lemons, 2017):

- y tends to a nonzero constant (this will likely be some inherently important value),
- y tends to some power law of the form cx^a, where c is a constant and a is a real number that cannot have dimensions,
- y oscillates in some periodic fashion, for example, such that $y(x) = y(nx)$, where n is an integer, and
- y does something else (e.g., tends to zero or infinity)

The last item on the list implies a huge unexplored complexity, but it is nevertheless useful to classify functions in this way.

For a function of two or more variables, the problem is more complicated—for example, given

$$z = f(x, y).$$

In any limit, we should consider the behavior of all of the parameters. For example, consideration of small values of x does not necessarily imply that $z \cong f(y)$. Consider that the function may involve a product of x and y. Consider also that the limit you impose on x may affect the limiting behavior of y. We consider this a bit further in an example problem at the end of this chapter.

2.7   Summary

- A wide range of quantities can be decomposed into primary dimensions for length, time, mass, temperature, and electric current: L, T, M, θ, and I.
- Dimensional analysis strives to reduce the complexity of problems by determining allowable relationships among variables based on an analysis of their basic dimensions.
- In this book, we will hypothesize functions for quantities of interest and then combine dimensional analysis and physical intuition and experimental observations to simplify and even evaluate such unknown functions.
- Concepts important to evaluating unknown functions include functional dependence on constants and ranges of applicability.
- The principle of dimensional homogeneity (PDH) requires that all terms in an equation must have the same dimension.
- The PDH is also useful in checking for errors and hypothesizing functions.

- Asymptotic behavior of functions is useful in classifying them or identifying important solutions and variables.

Problems

2.1. Consider the increase in pressure felt by a cupped hand extended out of car travel-ing at a speed of u. We shall see in the next chapter that the maximum pressure that can be generated (near the center of the cupped hand) can be estimated as follows:

$$p_0 = p + \frac{\rho u^2}{2},$$

where p is the local pressure of the air (in a frame of reference moving with the air) and ρ is the mass density of the air. Express this three-term equation as an unknown function with p_0 as the dependent variable and all other variables as independent dimensional variables. Use similar notation to show an approximation for this func-tion for the limit where the velocity tends to small values.

2.2. Pose several plausible relations (i.e., with the correct dimensions) for the volume of an oblate spheroid as a function of its semiaxis a and the distance c from its center to the pole (along axis of symmetry). Check the various limits of your equations and eliminate all but one. Check if the case of $c=a$ correctly yields the equation for the volume of a sphere.

2.3. The equation for the time t response $x(t)$ of an underdamped harmonic oscilla-tor with a damping coefficient γ, a natural frequency ω, and a phase α can be written as

$$x(t) = x_0 e^{-\gamma t} \cos(\omega t - \alpha),$$

where x_0 is the initial value of the response variable. This same basic formulation is used to model a wide variety of physical phenomena including a simple (low-angle displacement) pendulum; a mass, linear spring, and dashpot system; and an electrical circuit composed for a resistor, capacitor, and inductor. In such systems, we could interpret x as the angle, position, or voltage, respectively. Using the notation described in this chapter, we can write this functional relationship as follows:

$$x = f(x_0, \gamma, t, \omega, \alpha).$$

   a. Use a computer to plot the function using the following numerical values: $\gamma = 0.1$, $\omega = 1$, and $\alpha = \pi/2$.

   b. Use either your plotting software or your knowledge of mathematics to evaluate the various limits. As we discussed in section 2.6, be on the lookout for various forms of the asymptotic limits of the variable x (assume $x_0 = 1V$ for all limits). For each case, plot or consider how the x varies with time t.

- $\alpha = 0$ and ω to small values while γ is finite
- $\omega = 1$, $\alpha = 0$, γ tending to small values.

c. Use the notation formulation to describe (i.e., summarize) each of these limits.

$$x = f(x_0, \gamma, t, \omega, \alpha).$$

3

Some Basic Physical Principles

Law II: The change of motion [time derivative of momentum] is proportional to the motive force impressed [external force], and is made in the direction of the right line in which that force is impressed [co-linear, a vector quantity].

—Isaac Newton, *Principia* (1687)

In this chapter we summarize a few physical laws and principles that we will invoke in examples. This material is readily available in undergraduate physics texts and an introductory fluid mechanics book. The discussions assume an introductory level understanding of Newton's laws of motion, conservation of energy, and definitions of basic quantities such as velocity vectors, mass, and momentum. Some basic understanding of fluid flow principles is useful, because fluid mechanics offers many of the examples of dimensional analysis.

3.1   Newton's Second Law

Newton's second law in its basic form applies to a defined system. A system is some predefined mass or collection of masses (e.g., an object or objects). We will define the sum mass of the system as m. For arbitrary density distributions within the system $\rho(x, y, z)$, we can express the scalar m as a definite integral over the system volume(s)

$$m = \int_{\text{sys}} \rho \, d\forall. \tag{3.1}$$

Now, the momentum of the system is a sum of the differential momentum of each infinitesimal part, and it is a vector quantity. For nonuniform density and nonuniform velocities within the system $\bar{V}(x, y, z, t)$, the system's momentum $\bar{p}$ can be expressed as a definite integral

$$\bar{p}_{\text{sys}}(t) = \int_{\text{sys}} \rho \bar{V} \, d\forall. \tag{3.2}$$

For the special case where velocity (relative to some coordinate system) is uniform throughout the system, we can write

$$\bar{p}_{sys}(t) = \bar{V}\int_{sys} \rho d\forall = m\bar{V}. \tag{3.3}$$

Newton's law is applicable to an inertial frame, and it relates the sum of all external forces $\bar{F}_{ext}$ acting on the system to its rate of change of momentum:

$$\sum \bar{F}_{ext} = \frac{d\bar{p}_{sys}}{dt}. \tag{3.4}$$

Equations (3.1), (3.2), and (3.4) represent Newton's second law and can be applied to collections of deformable bodies and fluids, from planets to dolphins and to a specific selection of a thousand raindrops. For the special case where the velocity throughout the system is uniform (translational motion of a rigid body) and mass constant (e.g., no relativistic effects), we combine 3.3 and 3.4 as

$$\sum \bar{F}_{ext} = m\frac{d\bar{V}}{dt} = m\bar{a},$$

where $\bar{a}$ is the acceleration of the system. Note that Newton's second law is a vector equation, so in three-dimensional space it implies three expressions in terms of the components of $\bar{F}_{ext}$ and $\bar{a}$. That is, $\sum F_{ext,x} = ma_x$; $\sum F_{ext,y} = ma_y$; and $\sum F_{ext,z} = ma_z$ must be satisfied independently.

3.2   Conservation of Energy in Dynamics

The law of conservation of energy can be stated as follows:

> In a closed system (a system isolated from the surroundings), the total energy of the system is conserved.

This seemingly simple law has wide applications in dynamics, thermodynamics, fluid flow, heat transfer, and so on. For our purposes, we shall consider here only two types of energy that are applicable to simple problems in dynamics. The scalar quantity kinetic energy (*KE*) of a system of mass m moving with velocity vector $\bar{V}$ can be expressed in terms of the square of the magnitude of this vector, $|\bar{V}|^2$.

$$KE = \frac{1}{2}m|\bar{V}|^2.$$

Changes in *KE* can be expressed in terms of changes in $|\bar{V}|$.

The potential energy associated with a system is an energy associated with its location within a conservative field. This location is measured relative to some reference point. For a simple mass in a uniform gravitational field, we can express the scalar potential energy *PE* as

$$PE = mgh,$$

where h is the height relative to the reference point. Changes in PE are related to changes in h. For some system undergoing a change from state 1 to state 2, and where the primary form of energy conversion is between kinetic and potential energy, we can write

$$KE_1 + PE_1 = KE_2 + PE_2, \tag{3.5}$$

or

$$\frac{mV_1^2}{2} + mgh_1 = \frac{mV_2^2}{2} + mgh_2, \tag{3.6}$$

where V is the magnitude of velocity.

3.3   Basic Concepts of Fluid Flow

A fluid is any substance that deforms continuously when subjected to a shear stress (i.e., stress acting parallel to a substance). A fluid will deform continuously regardless of the magnitude of the stress. Two points to consider:

- Liquids and gases are fluids.
- The word "continuously" is important. A solid deforms under shear stress but, as a long as the elastic limit of the solid is not exceeded, it does not continue to deform.

Viscosity is used to quantify the friction within a deforming fluid. In a formal hypothesis, Newton defined it as "the resistance arising from the want of lubricity in the parts of a fluid" (see Motte's 1850 translation). It provides a measure of stresses associated with gradients in the velocity of the fluid. We here shall deal with typical fluids such as air, oil, and water, which satisfy Newton's law of fluids. For Newtonian fluids, the shear stress τ (force per unit area) within a fluid (or at a fluid/solid interface) is directly proportional to the local velocity field gradient. In a velocity field varying in only one dimension y, we can express this physical law as follows

$$\tau \propto \frac{du}{dy}, \text{ Newton's law of fluids}$$

where $\propto$ indicates proportionality. The constant of proportionality is called the dynamic viscosity μ. A Newtonian fluid is therefore defined as one that obeys the following relationship:

$$\tau = \mu \frac{du}{dy}. \tag{3.7}$$

Figure 3.1 helps explain this relation. To a very high degree of accuracy, the fluid immediately touching a solid wall moves at the speed of the wall; and this is called the "no-slip condition" (fluids do not "slip" relative to the wall). The fluid moves near the wall by deforming. In this example, the horizontal component of velocity u deforms only along

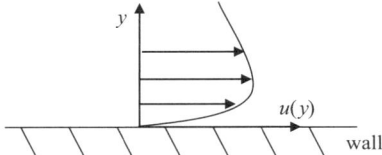

Figure 3.1
A plot of an arbitrary velocity profile of a fluid moving near a solid wall (wall is at $y=0$). The fluid velocity at the wall is zero as per the no-slip condition. The velocity of the fluid varies continuously as we get further from the wall. The gradients create stresses. The three horizontal arrows above the abscissa are simply a visual aid showing variations in velocity.

the normal direction y, so we write some arbitrary function $u(y)$. The local stress is a force per unit area in the planes parallel to the wall, and acts to resist the fluid motion. The shear stress is analogous to friction between two solids slipping with respect to one another (but here caused by continuous velocity gradients). The dimensions of viscosity can be determined from the relation above. For example,

$$\mu = \frac{\tau}{du/dy} \text{ [dimension of ML}^{-1}\text{T}^{-1} \text{ (kg·m}^{-1}\text{s}^{-1} \text{ in SI units)].} \tag{3.8}$$

The ratio of dynamic viscosity to density of a fluid often arises in analyses, and it is called the kinematic viscosity. This ratio is given a special symbol, v, so that

$$v = \frac{\mu}{\rho}, \tag{3.9}$$

which has dimensions of L^2/T (m²/s in SI units) and can be interpreted as quantifying the diffusion of momentum (e.g., in a gas due to the thermal motion of molecules) along the velocity gradient.

Last, although the details are beyond the scope of this book, the flux (the flow) of momentum through some area A can be expressed in terms of two velocities of the fluid: The local velocity relative to the area (e.g., the velocity that determines flow rate crossing the surface), and the velocity with respect to some inertial reference frame that determines the fluid's momentum and appears in Newton's second law (see White, 2016). For a one-dimensional flow at velocity V perpendicularly to some area A, we can write

$$\text{flow of momentum} = \rho V^2 A = \text{dimension of force.} \tag{3.10}$$

Indeed, changes of this quantity indicate reaction forces associated with inertia.

In fluids, the grouping ρV^2 is common. For example, students of fluid mechanics use the Bernoulli equation valid for steady, incompressible fluid flow (White, 2016) to estimate the pressure rise associated with frictionless deceleration of a moving fluid stream. Think here of your cupped hand catching the wind outside a speeding car. The pressure built up in your hand is associated with a deceleration of the fluid. For an approximately frictionless deceleration of the fluid of density ρ from velocity V to zero, the rise in pressure can be quantified as the dynamic pressure and expressed as:

$$\text{dynamic pressure} = \frac{\rho V^2}{2}.$$

The force associated with this pressure rise (this stagnation of the fluid upon) area A is then a force equal to the change in momentum of that stream.

$$\text{maximum force imparted by dynamic pressure on an area} = \frac{\rho V^2 A}{2}.$$

We see that it has the same dimensions of and a very similar formulation to the flow of momentum, since these expressions are closely related.

Laminar versus turbulent fluid flow: Last, we will distinguish between two regimes of fluid motion. In laminar motion, the fluid moves within layers or "laminae" and this is characterized by a steady, smooth motion of the fluid. If driven so, the motion may be complex and even three dimensional, but it is characterized by a low value of inertial forces (the characteristic inertia of the flow, which has dimensions of force) compared to viscous forces. In laminar flow the primary balance of forces is often between viscous and pressure forces. A fluid flow is very much laminar and damped by effects of viscosity if you remove the forcing function (e.g., applied pressure difference or the motion of a solid body), and the fluid quickly comes to rest due to viscous forces that dampen the motion. Laminar flows are analogous to overdamped dynamic systems.

Turbulent fluid flow is characterized by a high ratio of inertial forces to viscous forces. In turbulent flow, all forces can be important including inertial forces, pressure forces, and viscous forces. For larger, low-viscosity flows traveling at high velocity, viscous forces can become less important relative to pressure and inertial forces. Turbulent flow is always chaotic, fluctuating, and three-dimensional (3D). A flow is turbulent if, after sudden removal of a forcing function, the fluid continues to deform and fluctuate in unsteady 3D motions (e.g., shedding 3D vortices that interact with each other).

Whether or not a certain fluid flow will be laminar or turbulent is related to its Reynolds number, Re, where

$$Re = \frac{\rho V L}{\mu}.$$

V and L are the characteristic velocity and length scales of the flow and

- Low Re implies laminar flow.
- High Re implies inertial effects and, if sufficiently high, turbulent flow.

The exact value of the "transition" Re dividing laminar and turbulent flow depends on the flow geometry. For example, a jet of fluid ejected into an otherwise stagnant, large body of fluid has one characteristic transition Re, while pipe flows have another.

We shall discuss Re in more detail throughout this book, and this will help show its relation to dynamic pressure and Newton's law of fluids.

3.4 Summary

- Newton's second law relates the external forces applied to a strictly specified system to the rate of change of the momentum of that system:

$$\sum \bar{F}_{\text{ext}} = \frac{d\bar{p}_{\text{sys}}}{dt}.$$

- System momentum is an instantaneous integral over each differential portion of the system, described as

$$\bar{p}_{\text{sys}}(t) = \int_{\text{sys}} \rho \bar{V} d\forall.$$

- For the special cases where the velocity throughout the system is uniform (translational motion of a rigid body) and mass constant:

$$\sum \bar{F}_{\text{ext}} = m\frac{d\bar{V}}{dt} = m\bar{a}.$$

- Conservation of energy states that total energy is conserved for a closed system. In such a system, changes in state describe where energy is converted from one type (or more than one type) to another type (or more than one type).

- In this book, we will restrict ourselves to simple cases of dynamics wherein we trade off between kinetic and potential energy. For such systems undergoing a process between states 1 and 2, we can write:

$$KE_1 + PE_1 = KE_2 + PE_2.$$

- Fluids are defined in terms of the response of the material (continuous deformation) to an applied shear stress. Liquids and gases are fluids.

- Viscosity for a simple Newtonian fluid is defined as the ratio of shear stress to the shear rate. In one-dimensional flow, shear rate is the spatial rate of change (slope) of velocity expressed as du/dy.

- To high degree of accuracy, fluid velocities at a solid surface equal the velocity of the surface. This is known as the no-slip condition.

- The rate of momentum carried by a fluid at velocity V and flowing perpendicularly through some area A is $\rho V^2 A$, which has dimension of force.

- Frictionless deceleration of an incompressible fluid results in a maximum increase of pressure equal to $\rho V^2/2$, known as the dynamic pressure.

- Reynolds number is used to quantify the relative importance of inertial forces and viscous courses in a fluid flow problem: $Re = \rho VL/\mu$.

- Low Re implies highly damped, laminar flows where viscous forces quickly dissipate momentum. High Re implies underdamped flows where momentum persists.

- Sufficiently high Re leads to highly chaotic, three-dimensional fluid motion known as turbulence.

- The Re separating laminar from turbulent flow is specific for each problem (each geometry, flow angle, and so on).

Problems

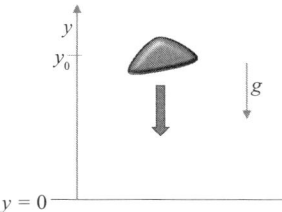

3.1.   Consider the velocity at which a rock hits the ground from some initial height $y = y_0$ (and under constant acceleration g). Neglect air resistance. Apply the first law of thermodynamics to derive an expression for the velocity V as a function of height y. What is the velocity when the rock hits the ground?

3.2.   Consider again the falling rock of the preceding problem. Again, neglect air resistance.

    a. Apply Newton's second law to describe the motion of the rock. Your description of the motion should include the velocity versus time and the position versus time (the functions $V(t)$ and $y(t)$, respectively). Hint: Apply Newton's second law to the rock and then integrate this expression in time. The initial state is $V(0)=0$ and $y(0) = y_0$.

    b. Use the solution of part (a) to solve for the velocity V as a function of position y. Check your answer by comparing it to problem 1.

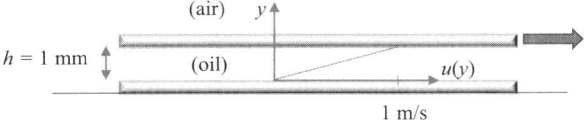

3.3.   Consider the following simple fluid flow problem. Two plates, each with an area of 1 m², are separated by a 1 mm gap filled with SAE 30 motor oil. The top plate is moving horizontally relative to the bottom plate and at a constant velocity of 1 m/s. Neglect the effect of the air.

a. Given the plot of y versus velocity in plot, give the value of the constant (and uniform) derivative du/dy. What are the dimensions of this derivative?

b. Find the viscous force (in Newtons) that the top plate imposes on the second plate. (Hint: First find the viscous stress on the top surface of the bottom plate).

4

Dimensional Analysis: Motivation and Introduction

> There could not be a language more universal … worthy of expressing the invariable relations of natural objects. … It is coextensive with nature itself; it defines all the sensible relations, measures the times, the spaces, the forces, the temperatures; this difficult science is formed slowly, but it retains all the principles it has once acquired.
>
> —Baron Jean-Baptiste Joseph Fourier (1822)

A system's mass and weight are each fundamentally different from its position. Acceleration is different from temperature. Momentum and kinetic energy are each different from heat capacity. So we tend to think of each variable as an independent, orthogonal quantity. We might mistakenly believe that, to understand the system, we must take into account the system's dependence on each variable separately. Dimensional analysis shows that this is a mistake that ignores the inherent ties between the system's variables. The variables are related in that they are derived from common ingredients. They are created from the primordial ooze of their primary dimensions. Consider that the dimensions of the vast majority of physical quantities are just combinations of mass, length, time, absolute temperature, and electric current.

The essence of dimensional analysis is this: We need not analyze the system's dependence on each of its salient variables, but rather on their interaction. The primary dimensions of the problem allow us to find their "blood ties" without further knowledge of the physics. We use these ties to create groupings of a problem's underlying variables, and these groupings can offer a drastic reduction of complexity. The result is a set of natural variables that are fewer and are independent of our choice of measurement system.

There are many uses for dimensional analysis, but here we will concentrate on demonstrating some of the most powerful. Our goals are to apply dimensional analysis to

- Present methodical ways of applying dimensional analysis
- Explore methods of combining dimensional analysis with physical intuition and experimental observations
- Analyze new phenomena and form testable hypotheses

- Design experiments to explore new phenomena
- Gain insight into the essential physics of the problem
- Obtain information from the fewest number of experiments, saving resources and time
- Present experimental data as compactly and as usefully as possible
- Demonstrate so-called collapse of experimental data
- Estimate the proper scaling laws that relate a model to a prototype
- Leverage results from older, similar systems to design new processes and machines

In the words of Ipsen (1960), our goal is to find the "natural variables," the nondimensional variables, from the "substantial variables" for physical problems.

4.1 Geometric Similarity

We shall initially and for most examples discuss dimensional analyses as applied to a single geometric shape at a time. That is, we will consider variations of length, velocity, density, mass, and so on, but the variation of length will typically be made so as to preserve the shape of the problem (e.g., the shape of the body, flow field, and the like). Similarly, we will preserve relative angles between velocities and objects (e.g., wind direction relative to a wing).

The concept of fixing a problem's shape and considering only changes to the absolute length scales of the shape is called geometric similarity. Idealized spheres are geometrically similar, while human beings are only approximately geometrically similar. A scale plastic model of a Boeing 747 jet with a 1 cm wingspan is geometrically similar to the actual, full-scale jet if it has the same shape. Once we "lock" into geometric similarity, we can fully describe the geometry by a single length scale (e.g., the wingspan or axial length).

Figure 4.1 shows two geometrically shaped airplanes: a 1:3 scaled-down model and a full-scale prototype. They share the same length ratios in all three directions (ideally including surface roughness). All angles are preserved. If we consider them being subject to flow, then the relative flow direction should be identical.

Geometric similarity is a good approach to learning about similarity. Eventually, we will consider approximate similarity of shapes (as in our examples of running animals or the splatter patterns of drops striking a surface). Sometimes such deviations require some knowledge of the physics of the problem to determine which aspects and/or lengths of the two "approximately similar" bodies should be matched.

In subsequent chapters, we will revisit the concept of geometric similarity together with kinematic and dynamic similarity. For now, we shall just say that we use geometric similarity and an analysis of the nondimensional parameters to ensure that we achieve dynamic similarity. The latter is the condition that the nondimensional ratios of geometry,

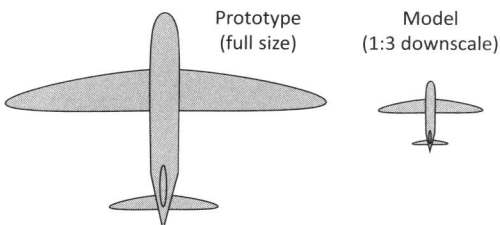

Prototype
(full size)

Model
(1:3 downscale)

Figure 4.1
Model and prototype of an airplane. The model and the prototype are geometrically similar, so all angles and
length ratios are the same.

velocities, and forces observed in the model are the same as the respective ratios of the
prototype. For example, such that the lift-to-drag ratio of the model be the same as lift-to-
drag ratio of the prototype. We will ensure dynamic similarity by matching the nondimensional parameters of the model with those of the prototype.

4.2   The Drawbacks of Brute Force Experimentation: We Need a Bigger Submarine

Before learning about the details, let us motivate our subject with an example application.
Imagine that you work for a small startup company interested in designing submarines.
At this point in the company's development, the basic shape of the submarine has been
fixed (e.g., by experience or tedious shape-optimization calculations). Before building a
full-scale prototype, however, the design and marketing folks would like to explore how
geometrically similar increases in scale (i.e., same shape, just larger) will affect the drag.
Note the drag and velocity of the submarine are related to the power required to propel it
(power dissipated by its motion is equal to the product of drag force and velocity). So this
question of size is tied to the requirements of the mission, range, payload, required engine,
fuel source, customer satisfaction, safety, and so forth. The company might want to find
out if it should consider, for example, thrice the scale of the submarine in order to fit three
passengers versus one.

Figure 4.2 shows a drawing of the submarine shape we will consider. We consider all
geometrically similar versions of this shape, so that we can specify all such submarines
with a single length scale. We arbitrarily choose its overall length b as shown.

We then postulate that the key variables that determine the drag force F_D of all submarines of this shape are as follows: length scale b, velocity V, fluid density ρ, and fluid
dynamic viscosity μ.

We wish to know the drag and hypothesize that the drag of this shape of submarine
when completely submerged is likely a function of four variables as follows:

$$F_D = f(b, V, \rho, \mu).$$

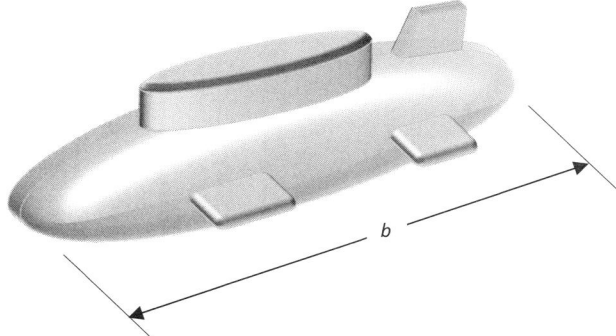

Figure 4.2
A submarine of length b. We consider all geometrically similar submarines, and so we fully describe the geometry with this length scale.

It is important to note that once we fix the shape of the submarine, the single length scale b (here, its length) completely describes every aspect of the geometry.

Brute Force Experimentation

First, we will describe a brute force method. We set out to find the functional relation between F_D and the independent variables. We study F_D vs. b vs. ρ vs. μ vs. V in all combinations. For simplicity, assume that the dependence of FD on each variable for constant values of the other variables requires at least 10 measurements. Hence, we will need on the order of $10^4 = 10,000$ measurement. Assuming 30 min per measurement and a low estimate of $20 per experimental data point (given cost of people, water or wind tunnel, building models, and such), we estimate

- Total cost of study: $200,000.00, and
- Total time: 2.5 yr at 8 h/day.

Using Dimensional Analysis

Using dimensional analysis (as described in the next chapter), we can vastly reduce time and cost. First, dimensional analysis will point out that we have asked the wrong questions regarding the variables. We do not need to know F_D vs. b, F_D vs. V, and so on. Instead we only need to know a single, new function F. This function describes the relationship between two new nondimensional parameters, two new "natural" variables:

$$\frac{F_D}{\rho V^2 b^2} = F\left(\frac{\rho V b}{\mu}\right).$$

If we identified the correct variables, all of the information of the five dimensional function $F_D = f(b, V, \rho, \mu)$ is compressed within the two dimensional space described by the function F. To evaluate this new function, we proceed as follows:

- Choose any single fluid, hence a single ρ and a single μ; for example, we can choose atmospheric air at room temperature
- Build a small-scale submarine; say a 10 cm long, 3D printed plastic model
- Measure 10 Vs for 10 F_Ds (10 experiments) and plot these data in terms of the non-dimensional variables associated with function F
- Use these 10 points to estimate the shape of the function F

Our new estimate has the following cost:

- Total cost of study: \$200.00
- Total time $= 5$ h

The process of getting from $F_D = f(b, V, \rho, \mu)$ to $\dfrac{F_D}{\rho V^2 b^2} = F\left(\dfrac{\rho V b}{\mu}\right)$ is a simple method we will describe in the next chapter. First, we will pause to consider some further consequences.

4.3   Comments on the Submarine Example

First, a word of caution: If we missed an important variable in identifying the independent variables of the originally hypothesized function f, then our analysis fails. In dimensional analysis, an important missing variable has dire consequences. We might find, for example, that some derived function would have multiple values for a single "condition" (we will discuss this further and analyze examples).

On the other hand, if we include a superfluous dependent variable in the function f, we will find that f has little or no dependence on this variable for our range of consideration—and this is a waste of experimental time and cost. For example, if we consider adding the speed of sound c to function f as follows

$$F_D = f(b, V, \rho, \mu, c).$$

Dimensional analysis would yield

$$\frac{F_D}{\rho V^2 b^2} = F\left(\frac{\rho V b}{\mu}, \frac{V}{c}\right),$$

where V/c is known as the Mach number Ma. Ma is relevant when the velocities approach the local speed of sound of the fluid. In analyzing a submarine designed to operate at typical submarine conditions (e.g., velocities of at most a few tens of meters per second in water), we would find that the drag term $F_D/\rho V^2 b^2$ is insensitive to the value of the ratio V/c if the ratio $\rho V b/\mu$ is held constant. Such an observation would require substantially more experiments but the weak dependence on Ma would eventually become clear. Consider that evaluation of the preceding three-dimensional relation would require variations of $\rho V b / \mu$ while holding V/c constant and vice versa. However, we would eventually realize our mistake but would have wasted time and money exploring the effect of the Mach number on typical submarine drags.

4.4   Drag Coefficient as a Tool to Reduce Complexity

The latter submarine example helps motivate dimensional analysis and reduction of complexity. Before we demonstrate how this is accomplished (see chapter 5), it is worth exploring three additional applications. These further emphasize the breadth of physics involved and the utility of reducing the complexity of something like shape-specific drag from six to two variables.

4.4.1   Effect of Streamlining

Recall that, in the submarine example, we derived a relation between two nondimensional parameters. These two parameters appear so often in fluid mechanics that they are given the following names:

$$Re = \frac{\rho V b}{\mu} = \text{Reynolds number}$$

$$C_D = \frac{F_D}{\frac{1}{2}\rho V^2 b^2} = \text{drag coefficient}$$

The Reynolds number is a famous nondimensional parameter that we will describe in more detail later. The "1/2" factor in the denominator of C_D is an arbitrary, traditional tribute to Euler, who derived equations where dynamic pressure $\rho V^2/2$ appears prominently (see section 3.3). Note also that the function F is arbitrary, so we can just think of a modification of the function hypothesized to include this "1/2" in the denominator. Note also that the length squared in the drag coefficient definition is often replaced with a specific area associated with the body in question. For example, for a cylinder oriented with its axis perpendicular to the flow, the area is taken as the product of diameter and cylinder length. For a wing in the shape of a solid of extrusion (uniform cross-section in one dimension) the area is taken as the so-called planform area of the wing. Judicious choices for the area in the formulation of a drag coefficient will be discussed later in section 10.3.

Figure 4.3 demonstrates the degree to which drag coefficients are shape dependent and the immense importance of streamlining to reduce drag. Figure 4.3a shows a smooth, low-drag, symmetric airfoil (wing shape) oriented directly into the flow (i.e., zero angle of attack). As a comparison, Figure 4.3b shows a smooth cylinder drawn such that its diameter d is equal to 110 times smaller than the airfoil's chord length, c (the distance between the leading and trailing edges of the airfoil). The airfoil shape is an accurate rendering of a National Advisory Committee for Aeronautics (NACA) airfoil number 63(3)-018. This

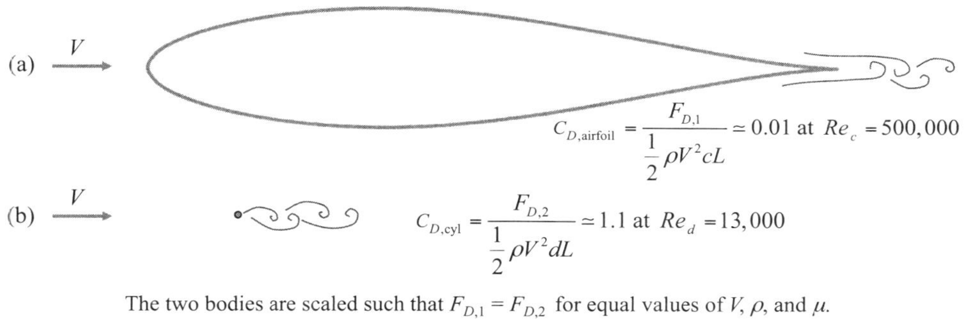

The two bodies are scaled such that $F_{D,1} = F_{D,2}$ for equal values of V, ρ, and μ.

Figure 4.3
A comparison of two bodies illustrating the importance of streamlining on fluid drag. Drawing (a) shows a cross-section of a smooth NACA 63(3)-018 airfoil at zero angle of attack. Here, the drag coefficient is equal to 0.01 (at Re = 500,000) and quantified in terms of a projected area equal to the product of chord length, c, and spanwise length, L. Drawing (b) shows a smooth cylinder subject flow at the same velocity. Here the drag coefficient is equal to 1.1 and is quantified in terms of the product of diameter and spanwise length, dL. The relative scale of the two figures is chosen such that the ratio of the airfoil chord length to the cylinder diameter is 110. At this scale, the two bodies would have the same absolute value of drag (e.g., measured in dimension of force) when subject to the same fluid velocity and fluid type.

symmetric shape is one of many airfoils designed and studied by and for NACA. NACA was a U.S. agency that was basically the predecessor of the National Aeronautics and Space Administration (NASA). See the website airfoiltools.com for drag and lift coefficient data for a large number of NACA airfoils as a function of angle of attack (another non-dimensional parameter) and Reynolds number.

The drag coefficient of this airfoil is 0.01 at a Reynolds number of $Re_c = Vc/v = 500{,}000$. As indicated in the figure, the drag coefficient is based on the planform area defined as the product of the chord length and a span L (into the page). Subject to the same fluid velocity and fluid type, the characteristic Reynolds number of the cylinder would be $Re_d = Vd/v = 13{,}000$, since $c/d = 110$. At this Reynolds number, the cylinder's drag coefficient is about 1.1 (Schlichting and Gersten, 2000). Now the cylinder's drag coefficient is defined in terms of the projected area dL. Hence, the figure was created so that each body has the same product of drag coefficient and characteristic area, that is, $C_{D,\text{cyl}}\, Ld = C_{D,\text{airfoil}}\, Lc$. For the relative scale drawn, the two bodies will exhibit the same value of drag (e.g., in Newtons) when subjected to flows of the same fluid at equal velocities. The influence of streamlining is profound. Consider that the airfoil has a cross-section that is 1,700 times larger than that of the cylinder.

4.4.2 Drag on a Smooth Sphere: A First Example of Data Collapse
One important reason to pursue dimensional analysis is to organize and rationalize a large amount of experimental data. Taking experimental data obtained by varying multiple

parameters and showing how the complexity (dimensionality) of the data is reduced when we plot on scaled coordinates is called "data collapse."

Consider the drag on a sphere with a smooth surface. Here, the concept of geometric similarity is very easy to apply. All smooth spheres are geometrically similar to each other and are characterized by a single dimension, their diameter d. We postulate that the key variables associated with determining the force of drag F_D on the sphere. We choose diameter d, velocity V, fluid density ρ, and fluid dynamic viscosity μ. These are basically the same variables we chose for geometrically similar submarines earlier (with diameter of sphere d replacing length b of submarine). Again, dimensional analysis leads to

$$\frac{F_D}{\frac{1}{2}\rho V^2 b^2} = f\left(\frac{\rho V d}{\mu}\right) \text{ or } C_D = F(Re).$$

Again, consider how much more compact is $C_D = F(Re)$ than $F_D = f(d, V, \rho, \mu)$ —two versus five variables, a significant reduction in complexity. Figure 4.4 helps to visualize the rich and various type of information incorporated within the more elegant function $C_D = F(Re)$. Figure 4.4a shows values of dimensional drag (in Newtons) versus velocity (in m/s) for a variety of sphere diameters and for flows in water, SAE motor oil, mercury, and air at room temperature and water at 90°C. These values are predictions obtained using expression given by Almedeij (2008) for drag on smooth spheres. Almedeij (2008) obtained the expression (equation 9 of that reference) by curve fitting a large number of published experimental data, and the data he analyzed were obtained by varying sphere diameters, velocities, fluid density, and fluid viscosity and measuring drag. Hence, these points are representative of experiments we could perform and the range of variation we would observe in changing all four of the independent parameters. Consider that we can measure drag and velocity to within about 5% uncertainty (see Almedeij, 2008).

Next, figure 4.4b shows a dramatic collapse of the exact same data when it is plotted in coordinates of C_D vs. Re. The seemingly complex, multidimensional behavior represented in part (a) collapses onto a single curve. This is an important simplification of complexity, and data collapse is a distinct benefit of dimensional analysis.

Consider the challenges with the wide variety of experiments required to obtain such data. The data points cover a Re range of 0.2 to 2E6, or 7 orders of magnitude. The fluid mechanics community has explored the relation $C_D = F(Re)$ for over 9 orders of magnitude changes in Re. For values of Re below about unity, Almedeij (2008) used an exact analytical solution known as Stokes solution, given by $C_D = 24/Re$. Duan (2015) shows similar data from seven experimental studies spanning decades of research and multiple laboratories in several countries (and a collapse of the experimental data onto a single curve). The curve of figure 4.4b has a complex shape, and there is no accurate theory starting from first principles for Re greater than about 5.

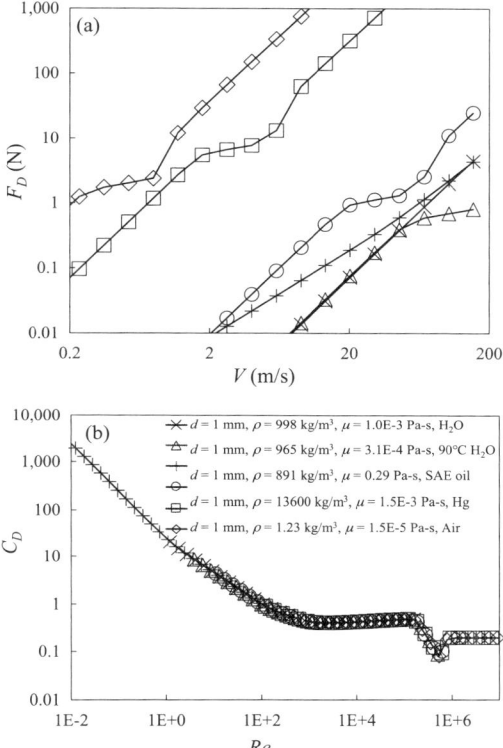

Figure 4.4
Collapse of drag values for smooth spheres. (a) Drag versus velocity for spheres of three diameters and three fluids: water, mercury, and air. The plotted points are values calculated from an analytical expression determined by Almedeij (2008) by fitting a wide variety of experimental data. Properties taken at 20°C unless indicated. (b) The same values of part (a) are here plotted as a nondimensional drag coefficient C_D versus a nondimensional Reynolds number, Re. The legend applies to both (a) and (b) and shows wide variation of parameters. The three fluids and two temperatures represent variations of density and dynamic viscosity. In the nondimensionalized space, the values collapse to a single curve. This is an example of data collapse obtained by reducing the number of variables in a problem.

This collapse of data for spheres of all sizes in various fluids and wide ranges of velocity can also be termed a *similarity* in the drag with a similarity relation expressed as $C_D = F(Re)$. In other words, we can interpret curves of figure 4.4a as representing an $n = 5$ variable solution in the five-variable space of F, d, V, ρ, and μ. Dimensional analysis shows these five variables can be projected onto only $n = 2$ variables: C_D and Re.

4.4.3 Drag and Terminal Velocity of a Skydiver
As a third example of the use and application of a drag coefficient, consider the physics associated with a skydiver. An anecdotal illustration of the relevant dynamics appears in

Figure 4.5
The drag force on a body is sufficient to cancel out the force of gravity. My body (on the right) is suspended in this "indoor skydiving" vertical wind tunnel in Fremont, CA (see http://sfbay.iflyworld.com).

figure 4.5, which shows my body suspended by drag force within a vertical wind tunnel with an air velocity of about 140 mph (63 m/s).

The analogous problem of a human falling through the air is shown schematically in figure 4.6. The figure shows a free-body diagram of a skydiver in free fall. We define a coordinate system with positive y-direction and positive forces upward. We use Newton's second law to relate the net forces on the body to its acceleration, the vertical y-component of which is

$$\Sigma F_{net,y} = F_D - mg = ma_y,\tag{4.1}$$

where a_y is the acceleration positive acceleration and g is the positive local value of gravitational acceleration. Before terminal velocity is achieved, a_y will be a nonnegligible, negative value (acceleration downward), so equation (4.1) becomes

$$F_D = m(g + a_y).\tag{4.2}$$

In the very early stages, F_D is initially small and the sum $(g + a_y)$ is nearly zero (a_y is $-g$). To relate to a drag coefficient, we simply and compactly cast the equation in terms of a drag coefficient. From the discussion of section 4.3, we know the drag force function can be characterized as

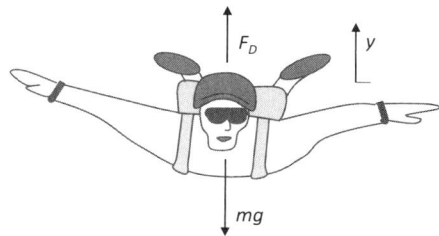

Figure 4.6
Free-body diagram of a skydiver in free fall at terminal velocity. We sum the forces acting about the centroid of the body. For zero acceleration, the drag force equals weight.

$$C_D = F(Re),$$ (4.3)

where

$$Re = \frac{\rho v \ell}{\mu} = \text{Reynolds number}$$ (4.4a)

$$C_D = \frac{F_D}{\frac{1}{2}\rho v^2 A} = \text{Drag coefficient}$$ (4.4b)

Here ℓ is the characteristic length scale for all geometrically similar bodies (in this spread-eagle configuration and falling belly down). Substituting equation (4.4b) into (4.2) and substituting $a_y = dV/dt$, where V is the vertical velocity:

$$\frac{1}{2} \cdot \frac{\rho v^2 A C_D}{m} - g = a_y$$

$$\frac{dv}{dt} = \frac{1}{2} \cdot \frac{\rho A C_D}{m} v^2 - g \qquad \text{where } v(t = 0) = 0$$ (4.5)

Since drag scales with the square of velocity, equation (4.5) is a nonlinear ordinary differential equation (ODE) and constitutes an initial value problem. The solution of this equation is given as a problem at the end of this chapter. For now, consider the steady state solution of equation (4.5), which describes the terminal velocity as

$$v = \sqrt{\frac{2mg}{\rho A C_D}}.$$ (4.6)

How do we estimate the drag coefficient? One way is to take typical values of bluff bodies and then match the projected area of the two bodies (which we will discuss in chapter 10). Here, we will leverage the data of Colino et al. (2014), who obtained estimate measurements of drag coefficients for Felix Baumgartner's record-breaking free-fall jump in a space suit from a balloon at an altitude of 40 km (very roughly halfway to space). The first stage of Baumgartner's initial drop was supersonic (for which our drag analysis is not applicable, since we neglected compressibility effects of the air flow and Mach number). For the subsonic (lower altitude) portion of his jump, Colino estimates $C_D \cong 1$ given our definition of drag coefficient and with Baumgartner in the typical (and stable) belly-down, back slightly arched position of figures 4.5 and 4.6. Interestingly, Colino's data showed the drag coefficient varies between about 1.0 and 1.1 during the last (low altitude) stages of the fall, helping to validate the assumption of approximately constant drag coefficient. They also provide a projected area estimate of about 0.25 m² (including Baumgarner's spacesuit). Taking air density as 1.2 kgm⁻³ (for the portion near sea level) and a skydiver of 72 kg mass, we can estimate $v \approx 68$ m/s = 150 mph.

4.5 Summary

- Geometric similarity considers a family of problems associated with a single specified shape but where absolute scale is allowed to vary. Geometric similarity includes preserving relative angles between velocities and objects (e.g., wind direction relative to a wing).
- Once we assume geometric similarity, the entire geometry can be fully described by specifying the value of a single dimension.
- In dimensional analysis, we hypothesize a functional dependence of a variable of interest. We then seek to reduce the complexity of the problem by combining dimensional variables into nondimensional variables.
- The nondimensional variables of a problem reduce the number of variables to consider. This reduction of variables implies a reduction in complexity. In real systems, this reduction in complexity (without sacrificing accuracy) can result in vast savings of cost and time.
- Nondimensional variables are the "natural" variables of the problem, capturing all of the underlying physics with a lower number of quantities (a lower dimensional space).
- Fluid flow drag on an immersed smooth body (of a specific geometry) is a good example of dimensional analysis that results in reduction of five variables into two nondimensional variables and a functional dependence as follows:

$C_D = F(Re).$

This compact, two-dimensional function (a single curve on a single plot), contains all of the intricate physics and underlying relationships among the variables F, b, V, ρ, and μ.

• We briefly reviewed several geometries to which we can apply (different) functions of the form $C_D = F(Re)$ and explored three examples of the ramifications and use of reducing the complexity of the drag problem.

Problems

4.1. Consider the nondimensional data of figure 4.4 for the drag of smooth spheres expressed as a drag coefficient as a function of a Reynolds number based on sphere diameter d: $C_D = F(Re)$. Consider the drag for a smooth sphere of diameter $d=1$ cm and traveling at a constant $V=2$ m/s in water at room temperature. Find the value of drag in Newtons. For these and other problems, you may find useful the thermophysical fluid property data tabulated in appendix A. Hint: First evaluate Reynolds number $\rho V d / \mu$.

4.2. Two geometrically similar submarines are tested in water. One is a small model with a length scale b that is 1/10th of the corresponding dimension of the full-sized prototype.

   a. The volume of the submarine of the small submarine is equal to $0.4b^3$. What is the volume of the large submarine?
   b. How much faster does the small submarine have to travel for it to have the same Reynolds number as the larger one? Consider the case where they travel through the same fluid (e.g., water) at the same temperature.

4.3. Consider now a more interesting use of figure 4.4b. Given ρ and μ of water at room temperature, use only figure 4.4b to estimate the diameter d of a smooth sphere that experiences a 100 N drag force while traveling at 2 m/s. Hint: This estimate requires an iterative solution, since there is no invertible function/fit to the curve in the figure. Start by guessing a value for d, then check your answer.

4.4. Solve for the dynamics as a skydiver accelerates and then reaches terminal velocity. Recall from section 4.4.3, we can write an ODE for these dynamics of the form:

$$\frac{dv}{dt} = \frac{1}{2} \cdot \frac{\rho A C_D}{m} v^2 - g.$$

Solve this equation numerically using a simple discretization as follows. Divide the (independent variable) time into small increments Δt (say 0.2 s intervals). To this end, write a small computer program to estimate the changes in velocity. Here are some notes on perhaps the simplest algorithm to get you started:

$v(0) = 0$ (initial velocity)
$t(0) = 0$ (time $= 0$)
$N = 800$ (total time)
$\Delta t = 0.2$ (0.2 s intervals)

for $i = 1: N$

$$t(i) = t(i-1) + \Delta t$$

$$\Delta v(i) = \left[\frac{1}{2} \cdot \frac{\rho A C_D}{m} \, v(i-1)^2 - g \right] \Delta t$$

$$v(i) = \Delta v(i) + v(i-1)$$

   end

This numerical solution method is known as Euler's explicit time advancement.

5

Dimensional Analysis Techniques

The square of the time taken by any planet to make a complete orbit is proportional to the cube of its mean distance from the sun.

—Kepler's third law (1619)

In this chapter, we first propose a set of rules useful in the execution of dimensional analyses of physical processes. These rules are part of a collection of rules that I have developed to help guide the selection of which variables to include and which to exclude. The proposed rules help hypothesize a function f of the dimensional variables form:

$$r = f(a, b, c, d, e, \ldots).$$

We will then use dimensional analysis to manipulate the variables of such a function to derive a function F in terms of relevant nondimensional variables

$$R = F(A, B, C, \ldots).$$

We shall see that the list A, B, C, ... is necessarily shorter than a, b, c, ... and so this will reduce the number of variables of the problem and help us seek elegant solutions.

In the second half of this chapter, we further describe dimensional analysis by analyzing some example problems and applying the aforementioned rules.

5.1 Rules of Thumb for Initial Hypothesized Function: What to Include or Exclude?

A goal of this text is to formalize the process of combining physical intuition and experimental observations with dimensional analysis. To this end, I have assembled a collection of useful rules of thumb that help structure such a hybrid analysis. The first five of these rules are presented in this section, and these deal with the most difficult and dangerous step of dimensional analysis: the initial selection of which variables to include and which to exclude. In chapter 6, we will propose and discuss six additional rules around simplifications of and approximations associated with functions of nondimensional variables.

A twelfth rule will be presented in chapter 11 (and table 11.1 summarizes the twelve rules) We shall refer to these rules as we apply them in the examples presented in this book.

The process of selecting variables for dimensional analysis takes the most experience. If you include too many variables, then the analysis yields less insight and can lead to significantly wasted time on experiments. In the next few chapters, we will present several examples of such overspecification. More importantly, missing one or more important variable in dimensional analysis is catastrophic. As we shall see, this can lead to unpredictable variations in data and poor understanding of the physics. As we shall also see, one ramification of excluding an important variable is that data cannot be "collapsed," as we saw in section 4.4.2. A second ramification is that we will observe (in experiments) variations in the value of a function even when the independent variables are held constant.

We begin with the first five proposed rules, which concern the selection of dimensional variables in the problem. These initial five rules are designated rules D1, D2, and so forth where the "D" denotes that they apply primarily to functions of dimensional variables.

5.1.1   Rule D1: Formulate in Terms of Known Algebraic Combinations

If a variable is known to appear in the problem exclusively within a certain algebraic combination, then you may hypothesize the combination as an independent variable. For example, consider some function of dimensional parameters as follows:

$$r = f(x, y, z).$$

If, for example, y and z have the same dimension, and you have good reason to believe they will occur only as a difference (and never independently), then you might consider modifying the hypothesis as follows:

$$r = g(x, y - z).$$

If you suspect that the formulation will involve both differences and absolute value of y and z, then write

$$r = f(x, y, y - z)$$

and manipulate the dimensional analysis such that the difference $(y - z)$ appears only in nondimensional terms where it is important and the parameter y appears in terms where it is important.

As an example, consider a simple analysis of the static displacement on a vertical linear spring (spring constant k with dimension of force per length). The spring is compressed vertically by a person holding a bucket of water. The human and bucket masses are m_h and m_b, respectively. We know the relevant dimension will be displacement x relative to the initial location of the end of the spring x_0. Given our experience with similar simple problems, it would be overly naïve to express the hypothesized function as follows:

$$x = f(x_0, m_h, m_b, k, g).$$

Instead, we might propose the following, which reduces the number of variables by two:

$$(x - x_0) = f(m_h + m_b, k, g).$$

If, for example, you are also certain that the variable will appear only as the product $k(x - x_0)$, then you can start with

$$k(x - x_0) = f(m_h + m_b, g).$$

A common example of such simplification is steady state heat transfer problems based on an imposed temperature difference. The heat transfer rate in such problems often scales as a temperature difference $T - T_0$ and not as absolute temperature.

5.1.2   Rule D2: Exclude a Variable Expressible as a Function of the Others

To decide whether to exclude a variable, it is useful to apply the following test: Exclude a variable if you can express it as a function of the others. Consider the following hypothesized function of dependent (dimensional) variable r:

$$r = f(x, y, z).$$

If by analysis or experimental observations (or even intuition) we can argue that parameter z is itself a function of only x and y (i.e., $z = g(x, y)$), then we can exclude z from our analysis and consider only

$$r = f(x, y).$$

We can do this even if the form of the function is not known. Note that this rule is equivalent to the "nested functions" concept described earlier in section 2.4.2.

5.1.3   Rule D3: Keep Dimensional Constants but Absorb Nondimensional Constants

In hypothesizing independent variables, you must include and retain physical constants that have units. Consider Albert Einstein's famous relation between energy E, mass m, and the speed of light in a vacuum c:

$$E = mc^2.$$

We can write this as

$$E = f(m, c).$$

However, just because the speed of light in a vacuum is a universal constant does not mean we can write or analyze the following:

$E = f(m)$. (incorrect)

The dimensional constant is fundamental to the relation. Energy cannot be formulated without both length and time dimensions. This is true for all constants with dimensions including universal constants such as Newton's gravitational constant, Boltzmann's constant, the charge of an electron, the electric permittivity of free space, Planck's constant, and so on.

On the other hand, we are free to exclude constants that are dimensionless, such as π or the exponent of unity. The latter dimensionless constants can be absorbed into the definition of the function you are exploring. Consider the equation for the area of a circle in terms of its radius r,

$$A = \pi r^2.$$

We can appropriately and without loss of generality write this as simply

$$A = f(r),$$

since both π and the squaring of r are part of the meaning of f.

5.1.4   Rule D4: Exclude Any Variable That Involves a Unique Dimension (No Blood from a Rock)

As per section 2.1 (and table 2.1), we consider describing variables in terms of basic dimensions; and in this book we consider L, T, M, θ, and I for dimensions of length, time, mass, temperature, and electric current, respectively. If one or more of these dimensions appear in just one variable (and not in others), then you may exclude that variable. Consider the following hypothesized function in terms of acceleration, velocity, length, and mass density

$$\underset{L/T^2}{s} = f(\underset{L/T}{x}, \underset{L}{y}, \underset{M/L^3}{z}).$$

If this list of variables is indeed sufficient to describe the phenomena, then we can immediately strike z from the list, since it is the only variable with dimension of mass. Consider that z cannot be made nondimensional except by dividing by itself. Division by itself yields unity, a dimensionless constant that can, by rule D3, be immediately be absorbed into the function definition. Hence, we can conclude that s must be described by

$$s = f(x, y).$$

5.1.5   Rule D5: If Invoked, Consistently Leverage Geometric Similarity

If you are applying geometric similarity, then all geometrical aspects of all scales of that shape are described by a single length scale. Hence, it is not appropriate to include other purely geometric descriptions. Consider some variable r as a function of a shape length s, area A, and volume V, as in

$r = f(s, A, V)$ (not needed under geometric similarity)

We should instead write $r = f(s)$, $r = f(A)$, or $r = f(V)$, since all aspects of the shape are derivable from any one of s, A, or V. Consider, for example, that any area can be expressed as $A = cs^2$, where c is a dimensionless constant for the selected shape. By rules D2 and D3, we can exclude A if we include s (and vice versa).

As a second example, including both density ρ and geometric length s precludes including a system mass m since $m = K\rho s^3$ where K is a dimensionless constant. Consistent with rules D2 and D3, this becomes superfluous under geometric similarity.

One important note on geometric similarity particularly relevant to many fluid mechanics problems is the effect of surface roughness. The roughness element length scale ε (i.e., the characteristic height of small "bumps" that characterize a body's surface roughness) can introduce a second important length scale to fully characterize the geometry of the body. This introduction then results in an additional nondimensional parameter known as the relative surface roughness, ε/d, where d is the (macroscopic) length scale used to characterize the absolute scale of the shape. This issue is discussed in more detail in chapter 8 (relative roughness in pipe flow problems) and chapter 10 (relative roughness in sphere drag coefficients).

5.2   Ipsen's Method: A Step-by-Step Process

In this book, I will emphasize Ipsen's method for dimensional analysis. The method is due to David Carl Ipsen (1960), who proposed a step-by-step method that is both simple and intuitive. The purpose of Ipsen's method is to derive a functional relationship in terms of nondimensional groups of variables. In chapter 8, we will discuss the more commonly taught method of the Buckingham Pi theorem and contrast it to Ipsen's method.

Ipsen's method is an elegant set of steps (see Ipsen, 1960, and White, 2016). In the following, I describe each step and then offers tips as to its application. I will later explain Ipsen's method more clearly using examples.

1. As with any dimensional analysis, identify the n key variables associated with the problem.

     Comment: Again, this is the most difficult step and takes the most experience. Consider carefully the rules at the beginning of this chapter.

2. Identify all fundamental/primary dimensions within the chosen n variables (e.g., M, L, T). The number of these is p.

3. Hypothesize some function of interest. To do so, select a dependent variable and $n-1$ independent variables. Express the primary dimensions for each variable.

    Brief example: For some drag force F of a specifically shaped body as a dependent variable, we might choose for independent variables a length scale b, velocity V, fluid density ρ, and viscosity μ. In this case, we write here

    $$F = f(b, V, \rho, \mu)$$

    We here assume geometric similarity (e.g., we consider only smooth bodies of the same shape, just different absolute scales) so we need a single length scale to describe all geometric aspects. Including both macroscopic shape and a relative surface roughness would require two length scales.

4. A good (but not always correct) guess for the number of nondimensional parameters we will produce is $n-p$.

5. Choose a primary dimension to eliminate from the formulation. We shall call this the current "chosen dimension."

6. Choose a variable containing this primary dimension, and call this the current "chosen variable." This should not be a variable you have chosen in a previous step 5, nor any variable that will reintroduce a dimension you have already eliminated (see the comment to step 8). You can also choose a variable or group of variables that is "simple to use" or that will arrange the terms in a manner you prefer.

    Comment: For example, say you choose M (mass) in step 5, then consider density as mass appears to the first power and is "easy to use." Dividing by density "removes" mass from the dimensions of the variable.

7. In this step, you will operate only on the remaining variables of the function, which contain the chosen dimension from step 5. For each remaining variable, raise the chosen variable (or group of variables) to some power m. Choose m such that multiplying or dividing each remaining variable by the chosen variable raised to power m will eliminate (cancel) the chosen dimension. For each of the remaining variables that contains the chosen dimension, choose a new value of m and multiple or divide as needed so as to cancel the chosen primary dimension. Do nothing with variables that do not contain the dimension chosen in step 5. (Note that leaving these variables alone is equivalent to a multiplication with the chose variable raised to the $m=0$ power). When you complete this step, the chosen dimension should not appear in the parameters of a newly derived function.[1]

---

1. Note that this step is consistent with the theorem of P. W. Bridgman (1931). Briefly, Bridgman's theorem states that the only type of functional operation that can operate directly on a quantity with dimensions is to raise that quantity to some power (and that this power is a pure number without dimensions). See Berberan-Santos and Pogliani (1999) for a detailed discussion and two proofs of the Bridgman theorem. Of course, functions such as logarithms, trigonometric functions, and exponentials cannot operate on arguments that have dimensions. At most, the arguments of such functional operators are composed of multiple dimensional variables that combine such that the combination is dimensionless (e.g., a ratio of two lengths).

Comment: If the chosen variable (or group of variables) is identical to one of the remaining variables, choose $m = 1$, then divide the variable into itself (equivalently, choose $m = -1$ and multiply). This operation yields unity. Unity is a pure number (exactly equal to the number one), and so this division eliminates the variable. It is now a dimensionless constant and, consistent with the aforementioned rule D3, it can be absorbed into the definition of the new function produced.

8. Choose a new primary dimension. Go to step 6 until you eliminate all primary dimensions.

Comment: Again, once you begin, you can use newly created groups for your choice in step 6 to eliminate primary dimensions. For example, you might choose a "group" in the form of a ratio of length and velocity of the form b/V (which has dimension of time T) to eliminate time. Choosing the simplest available group will save algebra and can help you guide which variables end up grouped with others.

In some cases, a poor choice in subsequent step 6s will "revive" a dimension you had previously eliminated. For example, say you have eliminated the dimension of time (e.g., using a frequency) in a previous step 7. If in a subsequent step 6 you choose velocity V to eliminate length, then division by velocity reintroduces time into the problem. You can avoid this by using groupings with the simplest dimensions. For example, given a choice, eliminate time with a previously identified ratio of the form b/V and not with V. Otherwise stated, choose variables for subsequent step 6s that do not contain the dimension of previous step 5s. It may sound confusing, but it will become clear with a little practice, and an example of this principle is discussed in detail in section 5.4.

5.3   Submarine Example Revisited

Let us again consider again the earlier submarine example, where we listed without derivation the nondimensional groups:

$$\frac{F}{\rho V^2 b^2}$$

$$\frac{\rho V b}{\mu}$$

Let's apply our eight steps of Ipsen's method.

1. Identify key variables: F, b, V, ρ, μ.

2. List primary dimensions M, L, T.

3. Hypothesize function of interest

$$
\begin{array}{ccccc}
F = & f(b, & V, & \rho, & \mu) \\
\text{MLT}^{-2} & \text{L} & \text{LT}^{-1} & \text{ML}^{-3} & \text{ML}^{-1}\text{T}^{-1}
\end{array}
$$
(a single length scale b as per rule D5).

4. Guess $n - p = 2$ nondimensional parameters.

5. Choose M.

6–8. Choose ρ to eliminate M, multiply or divided chosen variable ρ as needed to the relevant remaining variables, which include dimension M.

a. $\dfrac{F}{\rho} = f_1\left(b, V, \underbrace{\dfrac{\rho}{\rho}}_{\cong 1}, \dfrac{\mu}{\rho V b}\right).$

$\quad L^4T^{-2} \qquad L \quad LT^{-1} \quad L^2T^{-1}$

b. Choose V to eliminate T.

$\dfrac{F}{\rho V^2} = f_2\left(b, \underbrace{\dfrac{V}{V}}_{\cong 1}, \dfrac{\mu}{\rho V b}\right).$

$\quad L^2 \qquad\quad L \qquad L$

c. Choose b to eliminate L.

$\dfrac{F}{\rho V^2 b^2} = G\left(\dfrac{\mu}{\rho V b}\right).$

We find two nondimensional groups. As we shall discuss when we present rule ND1 in the next chapter, we can take the inverse of the argument of G and absorb this mathematical operation (inversion) into the definition of a new unknown function so that:

$$\frac{F}{\rho V^2 b^2} = F\left(\frac{\rho V b}{\mu}\right), \tag{5.1}$$

where

$$C_D = \frac{F}{\rho V^2 b^2} = \frac{\text{drag force}}{2 \cdot \text{dynamic pressure} \cdot \text{area}} = \text{drag coefficient.}$$

$$Re = \frac{\rho V b}{\mu} = \text{Reynolds number.}$$

The physical interpretation of these groups are as follows. C_D is the drag force normalized by a product of twice the dynamic pressure, ρV^2, and some area expressed as b^2. The Reynolds number is a measure of the inertial forces in the fluid flow divided by the viscous forces. For example, consider

$$Re = \frac{\rho V b}{\mu} = \frac{\rho V^2 b^2}{\mu \left(\dfrac{V}{b} \right) b^2} = \frac{\text{inertial force}}{\text{viscous force}}.$$

From equation (3.10) and its associated discussion, we recognize also that $\rho V^2 b^2$ is momentum flow through an area (changes of which imply inertial effects). The term $\mu(V/b)$ is a linearization of a one-dimensional viscous stress $\mu(dV/dy)$ (see section 3.3).

5.4   An Inelegant Application of Ipsen's Method

Briefly, consider a less successful application of Ipsen's method, where we ignore the comment to step 8. Consider again steps 6–8 for the latter problem of drag force. Say we begin with a reasonable choice of eliminating dimension L using length b:

1–4.   Same as above.

     5.   Choose to eliminate L.

6–8.

     a. Choose b to eliminate L.

$$\underset{\text{MT}^{-2}}{\frac{F}{b}} = f_1 \left(\frac{b}{b}, \underset{\text{T}^{-1}}{\frac{V}{b}}, \underset{\text{M}}{\rho b^3}, \underset{\text{MT}^{-1}}{\mu b} \right).$$

Now, we ignore the comment in step 8 and do not choose the simplest grouping to remove the next dimension M. Instead, we choose density (which contains length):

     b. Choose ρ to eliminate M.

$$\underset{\text{L}^3\text{T}^{-2}}{\frac{F}{\rho b}} = f_2 \left(\underset{\text{T}^{-1}}{\frac{V}{b}}, \underset{\text{L}}{b}, \underset{\text{L}^3\text{T}^{-1}}{\frac{\mu b}{\rho}} \right).$$

Our poor choice has brought length L back into the problem! We can recover by thereafter adhering to our rule of using the current simplest available variable.

     c. Choose b to eliminate L.

$$\underset{\text{T}^{-2}}{\frac{F}{\rho b^4}} = f_3 \left(\underset{\text{T}^{-1}}{\frac{V}{b}}, \underset{\equiv 1}{\underbrace{\frac{b}{b}}}, \underset{\text{T}^{-1}}{\frac{\mu}{\rho b^2}} \right).$$

    d. Choose b/V to eliminate T.

$$\frac{F}{\rho V^2 b^2} = F\left(\frac{\mu}{\rho V b}\right).$$

Ipsen's method can be applied skillfully to both reduce algebraic steps and guide the result.

5.5 Time for a Stone to Drop: Experimental Closure and Collapse of Data

We begin gently with a simple problem: the time for a stone to fall from some height, as depicted in figure 5.1. We purposely consider a case where we know the solution to the dynamics, but then pretend that we do not know the solution to see how far dimensional analysis will take us. We will also use this example to demonstrate the following two principles:

- Dimensional analysis can be supplemented by experiments to find the solution of the problem without knowledge of the equations that govern the underlying dynamics.
- Dimensional analysis collapses experimental data and enables us to present data in a simple compact form using a minimum of plots.

    First, let us remind ourselves of the known, underlying solution. Neglecting drag, the time t_0 for the stone to accelerate from rest and drop from initial height y_0 above the surface of Earth can be obtained from two integrations in time of Newton's second law for uniform and constant acceleration:

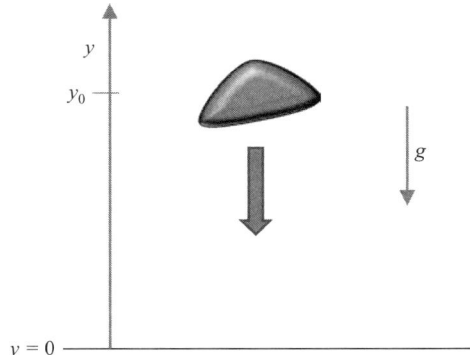

Figure 5.1
Simple dynamics problem of dropping a stone of mass m over a distance y_0. Neglecting drag, we assume constant acceleration g.

$$y(t) = \frac{-gt^2}{2} + y_0.$$

We leave the derivation of this equation as an exercise to the reader. We can solve this for the time for the rock to drop to Earth's surface by setting the function $y(t)$ equal to zero and inverting the expression as

$$t_0 = \sqrt{2y_0 / g}. \tag{5.2}$$

Next, we pretend that we do not know the solution and turn to dimensional analysis. First, we might hypothesize that the relevant variables are as follows:

Variable	Definition	Physical quantity
t_0	time for a stone to fall (the dependent variable)	T
h	initial height of the stone	L
g	acceleration of Earth's gravity (near surface)	LT^{-2}
m	mass of stone	M

We here purposely include a naïve selection of mass. Not knowing enough about the kinematics of falling bodies, we might indeed choose mass. Following Ipsen's method, this selection of variables yields the following hypothesized function and primary dimensions of each variable:

$$t_0 = f_1(h, \quad g, \quad m)$$
$$ \text{L} \quad LT^{-2} \quad \text{M}$$

Immediately, dimensional analysis provides us with a powerful insight. Assuming the list has at least as many variables as needed, we see that only a single variable has dimension of mass M. How can this be? We must conclude that mass cannot be relevant to the problem. Consider the proposition of dimensional analysis: that we should be able to reduce the functional expression into a set of nondimensional groups. How are we to "reduce" mass dimension M? Only by dividing mass m by itself, which yields unity—a dimensionless constant. We conclude that the other variables cannot be a function of mass, and mass cannot appear in the final dimensionless groups as per rule D4.

In light of this, we revise our hypothesis as follows:

$$t_0 = f_2(y_0, \quad g)$$
$$\text{T} \text{L} \quad LT^{-2}$$

Let's then apply Ipsen's method. Choose to eliminate L, using h.

$$t_0 = f_2\left(1, \frac{g}{y_0}\right).$$
$$\text{T} \qquad\qquad\quad \text{T}^{-2}$$

Choose t_0 to eliminate T.

$$1 = F\left(\frac{gt_0^2}{y_0}\right).$$

Now, a function that is always equal to a constant (here unity) must itself depend on a constant. We will describe this idea specifically as part of rule ND1, which we shall present in the next chapter. Hence, gt_0^2/y_0 must itself be a constant, so $y_0/gt_0^2 = c_0$, where c_0 is an unknown (dimensionless) constant. We note that this is clearly as far as dimensional analysis alone can take us. Dimensional analysis can in no way help us estimate or evaluate this constant.

5.5.1 Supplementing Dimensional Analysis with Experiments

Next, let us turn to supplementing dimensional analysis with experimental observations. Figure 5.2 shows experimental data I obtained by repeatedly dropping a 7.6 cm diameter ball.[2] The data of figure 5.2a are raw y-location measurements obtained by me via quantitative analysis of images of ball drops from heights of $y_0 = 2.26$, 1.68, 1.12 0.56, and 0.28 m, as shown in figure 5.2a. Briefly, the ball was dropped three times from each height, and images were quantitatively analyzed using a custom Matlab (Mathworks) script that automatically identified and tracked ball image centroids. Figure 5.2a shows raw experimental data of 365 successfully identified and tracked ball image locations, y, as a function of time, t, for the 15 experiments. The fairly high degree of spatial

---

2. Unpublished experiments quantifying a ball falling from heights of 2.26, 1.68, 1.12, 0.56, and 0.28 m. The plastic ball was 7.6 cm in diameter, had a mass of 153 g, and was fitted with a small green light–emitting diode for ease of tracking. The phone camera was approximately 3 m away to lower magnification (and hence ball image velocity at image plane). Videos were recorded under dim room light conditions in my living room using a Samsung Galaxy S8 phone (model SM-G950U) at frame rate of 59 fps. Images from movies were read into and analyzed using a custom Matlab (Mathworks) script. Briefly, color images were converted to grayscale, and a time-ensemble median image was subtracted from each frame to remove background (i.e., nonmoving image features). Image intensities were thresholded and resulting shapes were processed morphologically and subject to size thresholds for identification. Location data in Figure 5.2a were raw computed centroids for each of three realizations (drops) per tested height. The uncertainty in ball location is approximately a ball radius (3.8 cm), and the temporal uncertainty in tracking was approximately 16 ms. The uncertainty bars in Figure 5.2b reflect the uncertainty on the dimensionless quantity estimated using a propagation of error analysis (Coleman and Steele, 1998). The higher uncertainty at the lowest heights are expected as the relative error of both location and time are greater. I gratefully acknowledge Philip D. Santiago for help with the experiments.

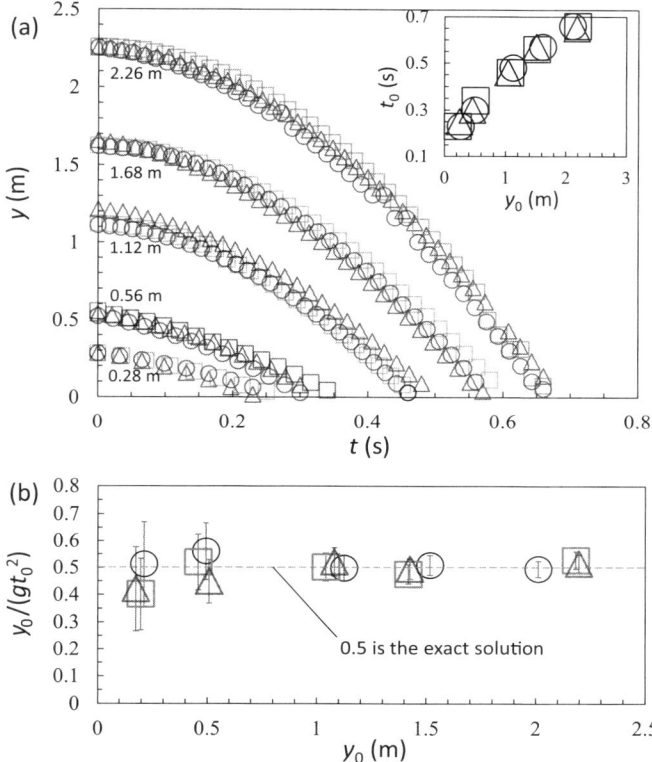

Figure 5.2
Experimental data for a falling body. The model "rock" for these data was a 7.6 cm ball tracked using automated image processing of videos obtained at 60 frames per second. (a) Each drop height was repeated three times, and the main plot shows raw experimental data of tracked and quantified centroids for ball images dropped from the five heights indicated by the labels. The inset shows 15 dimensional measured total drop times for the 15 experiments. (b) shows a collapse of the same experimental data of the inset of (a). The collapse of data is achieved by plotting the measurements in terms of the dimensionless parameter derived using dimensional analysis, gt_0^2/y_0 as a function dimensional initial height y_0. For reference, the exact theoretical value of this parameter (1/2) is shown as a dashed gray line.

and temporal resolution obtained from 59 images per second data are a bit of overkill for our purposes, but the data helps us visualize the parabolic dependence (on time) that was predicted by dimensional analysis. The shapes of the curves also hint at the underlying similarity in the data (note how the first ~0.25 s of each curve has the same curvature and final slope).

The type of data we need to supplement our dimensional analysis is shown in the inset to figure 5.2a. The inset plot shows the measured total drop time t_0 as a function of initial height y_0 for the 15 drops from five initial heights. Collectively, all of the *dimensional* data of figure 5.2 demonstrates the wide scatter of drop heights versus time for the problem.

Figure 5.2b helps show the utility and power of dimensional analysis. Plotted are again the raw experimental data for total drop times and initial heights (as per our initial analysis) but here in terms of the dimensionless quantity that we derived from dimensional analysis, y_0/gt_0^2. Note that there is a single value of this quantity for all such drops (i.e., for all heights). For reference, we here plot this quantity as a function of the dimensional drop height y_0 on the abscissa. The uncertainty bars in figure 5.2b reflect an uncertainty on the measured dimensionless quantity y_0/gt_0^2 and were estimated using a propagation of error analysis (see footnote to section 5.5.1). The greater uncertainty in y_0/gt_0^2 for small drop heights is due to the greater relative uncertainty associated with small-displacement, short-time drops. Taking a simple arithmetic average of the value of y_0/gt_0^2 for the nine drops from the three largest (most accurate) height experiments, we obtain a *measurement* for the constant predicted by dimensional analysis c_0

$$y_o / gt_o^2 = c_o \simeq 0.50 \pm 0.01 \text{ (value from experiments).}$$

The uncertainty bounds in this estimate of c_0 were calculated using a student t distribution for eight degrees of freedom (nine drops of the ball) and 95% confidence intervals. Compare this approximate result obtained by analyzing measurements of our derived nondimensional quantity y_0/gt_0^2 to the exact solution: From equation (5.2), we see that

$$y_0/gt_0^2 = c_0 = \frac{1}{2} \text{ (exact solution).}$$

The estimate obtained from experiments, 0.50 ± 0.01, is within 2% of the expected solution.

The physical process we explored here is rather simple and its solution known. However, our approach is a good example of how to combine dimensional analysis and experimental observations to complete (i.e., close) the analysis and, at least in this case, obtain a highly accurate solution. The collapse of data also shows the value of dimensional analysis in presenting data in a collapsed form comparable across bodies, realizations, observers, absolute scales, planets, and so forth provided the accuracy of the assumptions hold. In this case, the primary assumptions were approximately constant and uniform acceleration and negligible effects of drag.

5.5.2   A Note on Guiding the Nondimensional Parameters

Last, we note that, if we prefer, we can avoid the slightly pesky format derived earlier where a hypothesized function as equal to unity; namely, $1 = F(gt_0^2/y_0)$. Recall from our description of Ipsen's method that we are free to choose from among newly derived groupings to eliminate remaining dimensions and can then raise these to some power m. Hence, starting from the penultimate step in Ipsen's method,

$$t_0 = f_2\left(1, \frac{g}{y_0}\right).$$

We can choose the ratio $h^{1/2}/g^{1/2}$ to eliminate T, hence

$$t_0\left(\frac{g}{y_0}\right)^{1/2} = G(1) = \text{constant},$$

where this constant is simply $\sqrt{c_0}$, and the result is equivalent to the one above.

5.6   From Stones and Earth to Planets and Stars

Consider an extension of the problem of section 5.4. We wish to quantify the time scale in terms of Newton's gravitational constant. Here, recall the advice presented in section 5.1 that dimensional analysis must necessarily include universal constants that have dimensions. We hypothesize the following:

Variable	Definition	Physical quantity
t_0	time for a stone to fall (the dependent variable)	T
h	initial height of the stone	L
r	distance between stone and Earth center	L
m	mass of stone	M
M	mass of the Earth	M
G	Gravitational constant	$L^3M^{-1}T^{-2}$

where r is now some distance between the center of the stone of mass m and the Earth of mass M. G is Newton's universal gravity constant, which has derived units of

$$G = 6.67384 \times 10^{-11}\,\text{m}^3 \cdot \text{kg}^{-1} \cdot \text{s}^{-2}.$$

Our hypothesized function is then

$$t_0 = f_1\left(\begin{array}{ccccc} h, & r, & G, & m, & M\end{array}\right).$$
$$\begin{array}{cccccc} \text{T} & \text{L} & \text{L} & L^3M^{-1}T^{-2} & \text{M} & \text{M}\end{array}$$

By rule D2 we exclude the surface gravitational acceleration, since it is determined from the other variables (we will expand on this further in section 5.6). Also, from rule D3, we must keep the dimensional constant G in the problem. Note that Newton's law of gravitation is formulated in terms of the masses of two bodies and the relative distance of their centers (and not their shapes). Hence, the shapes of the bodies is not reflected in the formulation, only the distances between "point masses."

So, using Earth's mass M to eliminate primary dimension M:

$$t_0 = f_2\left(h, \quad r, \quad GM, \quad \frac{m}{M}, \quad 1\right).$$

T   L   L $L^3 T^{-2}$

Using r to eliminate L

$$t_0 = f_3\left(\frac{h}{r}, \quad 1, \quad \frac{GM}{r^3}, \quad \frac{m}{M}, \quad 1\right).$$

T T^{-2}

Using t_0 to eliminate T,

$$1 = F_1\left(\frac{h}{r}, \quad 1, \quad \frac{GMt_0^2}{r^3}, \quad \frac{m}{M}, \quad 1\right).$$

This gives us three nondimensional groups. Our insight is not as simple as in the previous section, but the analysis is now applicable to gravitational interaction between any two bodies in space, including celestial bodies. For example, consider the problem of the previous section but change "stone" to "planet" (or satellite) and "Earth" to "star." Here we can interpret time t_0 as a period (a cycle time, such as an orbit). In this celestial picture, the distances are so great, and the mass difference are so large, that we can simplify the problem with some physical insight.

Physical insight: We consider the mass M to be so much more massive than m (i.e., the star so much larger than the planet) that we can assume the dimensionless parameter m/M is vanishingly small. Similarly, we assume the center-to-center distances are vast, so that $h/r \cong 0$. As we discussed earlier in the chapter, this does not necessarily ensure that we can discard these variables. However, in this case, we can leverage our physical insight to conclude that the time scale t_0 in such a limit will indeed be governed solely by G, the mass of the large body, and the characteristic distance between the centers of the small and large bodies r. Note also the third parameter GMt_0^2/r^3 remains finite as the other terms vanish. Given these insights, we hypothesize

$$1 = F_2\left(\frac{GMt_0^2}{r^3}\right),$$

which implies (from rule ND1; see chapter 6) that

$$\frac{GMt_0^2}{r^3} = c_0, \quad \text{where } c_0 \text{ is some constant.}$$

Rearranging this, we derive a formulation of Kepler (circa 1610):

$$t_0 = \sqrt{c_0}\,\frac{r^{3/2}}{\sqrt{GM}}\quad \text{Kepler's third law } (\sqrt{c_0} = 2\pi).$$

Interestingly, we derive Kepler's third law describing the orbital period t_0 of a satellite around a much larger celestial body. Since our analysis has included only a single length scale r and no other geometric description, we might guess (correctly) that our result applies to circular orbits of bodies around relatively heavy bodies.

In Kepler's law, r turns out to be the semimajor axis of the elliptical orbit (the arithmetic average of the minimum and maximum distances between the two bodies). Dimensional analysis does not yield the latter definition of r, but it does suggest that there is such a distance to describe the period. Figure 5.3a shows a plot the semimajor axis of the orbits of planets in our solar system as a function of orbital period (Earth is by definition $y_0 = 1$ yr). The diameter

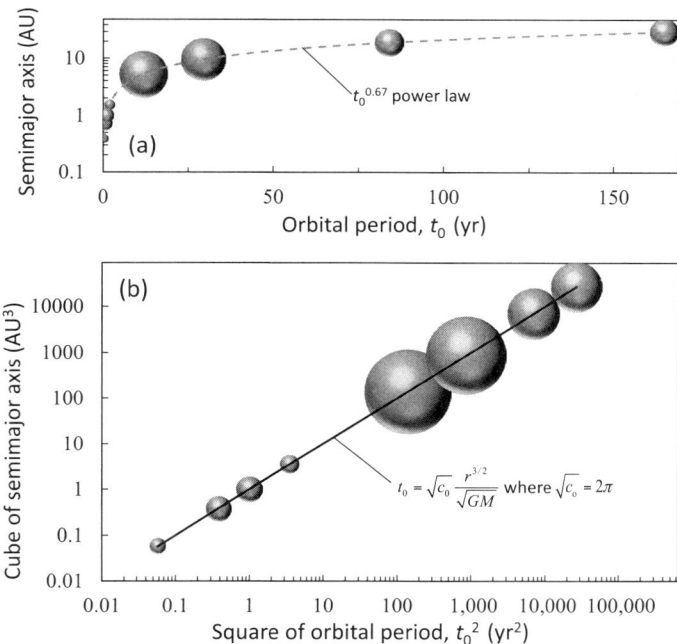

Figure 5.3
Scaling analysis for Kepler's third law. The top figure shows dimensional plots of the semimajor axis of the elliptical orbit versus orbital periods of planets (including dwarf planet Pluto) around the sun. The bottom figure shows the same data plotted as the cube of the semimajor axis r^3 versus the square of the orbital period t_0^2, as suggested by the scaling relation $r^3 = $ constant $(t_0^2 GM)$. The units are years (yr) for period and astronomical unit (AU) for distance (roughly the distance between Earth and the sun). Data point symbol diameters are proportional to planet projected area.

of the data symbols reflects the value of the square of the diameter of the body. Figure 5.3b shows how these values collapse nicely into a straight line in coordinates predicted by the dimensional analysis. A fit of this line provides the value of the constant to high accuracy.

5.7   Ignoring the Rule That We Exclude Determined Parameters: Including Both G and g

Let's look again at the dynamics of a mass from some height, but let's ignore rule D2, which dictates that we should include g or G but not both (since you can derive g from G, M, and r). We consider the problem specifically applied to motion near the Earth's surface. We start with the following function:

$$t_0 = f_1\left(h, \quad r_E, \quad G, \quad\quad m, \quad M_E, \quad g\right),$$
$$\text{T} \quad\quad \text{L} \quad\ \text{L} \quad \text{L}^3\text{M}^{-1}\text{T}^{-2} \quad \text{M} \quad\ \text{M} \quad \text{LT}^{-2}$$

where M_E is now specifically the mass of the Earth and r_E is Earth's radius. Applying Ipsen's method, we can derive

$$1 = F_1\left(\frac{h}{r_E}, \frac{GM_E t_0^2}{r_E^3}, \frac{m}{M}, \frac{gr_E^2}{GM_E}\right).$$

The result is correct but superfluously including both g and G results in four variables. In this case, we see that we have a nondimensional parameter, which is itself a nondimensional universal constant. Namely,

$$\frac{gr_E^2}{GM_E} \cong \text{constant}.$$

We can absorb such a dimensional constant into the function and derive

$$1 = F_2\left(\frac{h}{r_E}, \frac{GM t_0^2}{r_E^3}, \frac{m}{M}\right).$$

We shall discuss absorbing a nondimensional constant as part of rule ND5, which we will describe in chapter 6. Alternately, we can multiply the second and third variables in function F_2 to eliminate G from the term involving t_0 and derive

$$1 = F_3\left(\frac{h}{r_E}, \frac{gt_0^2}{h}, \frac{m}{M}, \frac{gr_E^2}{GM_E}\right).$$

We shall discuss reorganizations of such a function by multiplying arbitrary powers of the (nondimensional) arguments as part of rule ND1 in chapter 6. Again noting that last dimensionless parameter is constant, we have

$$1 = F_4\left(\frac{h}{r_E}, \frac{gt_0^2}{h}, \frac{m}{M}\right) \text{ or } \frac{gt_0^2}{h} = F_5\left(\frac{h}{r_E}, \frac{m}{M}\right).$$

Again, rewriting function F_4 as a new function F_5 will be discussed as part of rule ND1 in the next chapter. In any case, this is as far as dimensional analysis can take us.

At this point, we might apply some physical intuition for the problem of a stone dropped a small distance near Earth's surface. Note that h and t_0 remain finite as we consider m/M_E and h/r_E becoming vanishingly small. Hence, in the limit of small m/M_E and h/r_E we can hypothesize that the time will only depend on the height h and propose:

$$1 = F_5\left(\frac{gt_0^2}{h}\right) \text{ or } \frac{gt_0^2}{h} = \text{constant}.$$

This is the result we derived in section 5.4.[3]

5.8   Period of a Pendulum for Any Angle: Experimental Closure and Collapse of Data

Consider the period of a pendulum. The analysis of this is also well established. For example, there is a simple solution available for the case of small displacements. We shall use dimensional analysis to explore the more complex case of pendulums oscillating at any angle. Figure 5.4 shows some details of the type of schematic used to relate the force components on a pendulum and analyze its motion. Note that this is not a free body diagram but merely a breakdown of the gravitational force components.

We shall not leverage Newton's law but will immediately proceed to dimensional analysis using Ipsen's method. We first postulate a list of parameters as follows:

---

3. Note an alternate function F_4 could have been derived as follows using dimensional analysis:

$$1 = F_4\left(\frac{h}{r_E}, \frac{gt_0^2}{r_E}, \frac{m}{M}\right) \text{ or } \frac{gt_0^2}{r_E} = F_5\left(\frac{h}{r_E}, \frac{m}{M}\right).$$

However, it not then appropriate to discard the parameters involving h and m and keep only gt_0^2/r_E. For one thing, we do not expect the rock's travel time t_0 for motions near the surface of the Earth to be governed only by the radius of the Earth and also be independent of the height dropped. Such an assertion does not conform with our experience and observations. Instead, our experience shows that t_0 is finite even for very small values of h/r_E. These considerations serve as a good introduction and motivation for rule ND6, which we will discuss at the beginning of chapter 6. For now, we note that we must consider how all nondimensional terms behave in some limit.

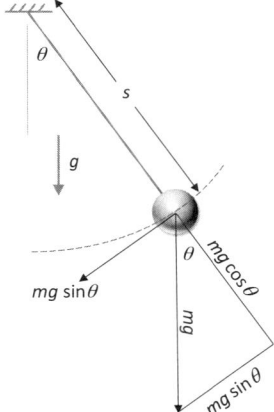

Figure 5.4
Typical schematic considered in the analysis of a pendulum with a small sphere of mass m, massless rod, frictionless pivot, length s, and sweeping over angle θ. We neglect drag and assume Earth's downward acceleration g.

Variable	Definition	Physical quantity
t_0	period of pendulum (dependent variable)	T
s	arm length	L
θ_0	initial angle	—
g	gravitational acceleration	LT^{-2}
m	mass of pendulum	M

The pendulum's very small weight is here approximated as a point mass, and so has no relevant length scale. We again hypothesize a function:

$$t_0 = f_2\big(\underset{\text{T}}{s},\ \underset{\text{L}}{g},\ \underset{\text{LT}^{-2}}{\theta_0},\ \underset{\text{M}}{m}\big).$$

As with the falling stone example, we again notice that the weight's mass is the only parameter that contains the primary dimension M. Hence, we can immediately conclude that the mass cannot be relevant here as per rule D2. So

$$t_0 = f_2\big(\underset{\text{T}}{s},\ \underset{\text{L}}{g},\ \underset{\text{LT}^{-2}}{\theta_0}\big).$$

Note the angle θ_0 is a nondimensional quantity and so already forms its own nondimensional group. Applying Ipsen's method to remove T with g and then to remove L with s, we have

$$\frac{gt_0^2}{s} = F(\theta_0).$$

Interestingly, this result is correct in that, for arbitrary initial angles, the period of a pendulum indeed depends on the initial angle! Larger initial angles yield longer periods. Only for small initial displacements does the period become insensitive to initial displacement. See Belendez et al. (2007) for exact solutions to the problem of large displacement pendulums where they quantify period in terms of incomplete elliptical integrals of the first kind.

Figure 5.5 highlights the importance of the nondimensional variables gt_0^2/s and θ_0. The top plot shows both experimental data and accurate predictions for the (dimensional) period of pendulums of lengths 1, 1.5, and 3.08 m. The data points are experimental data from two different sources: Experiment 1 (circles) data were obtained by Fulcher and Davis (1976) using a 3.08 m long piano wire and plumb bob. Experiment 2 (squares) data were obtained by Lima and Arun (2006) using a rigid 1.5 m long pendulum arm on a conical bearing and with electronic timing. The curves are predictions based on gravitational accelerations consistent with the gravitational acceleration magnitudes near the surfaces of Earth, the Moon, and Mars. These curves are based on an approximate solution to the large-displacement pendulum period problem, and have an error (relative to the exact solution) of less than about 0.1% for $\theta_0 < 69°$ and about 0.4% error near $\theta_0 < 90°$ (see Lima and Arun, 2006). The bottom plot shows the experimental data (obtained decades apart on different continents and using very different experimental setups) and the predictions plotted in the variables that we identified using dimensional analysis: gt_0^2/s and θ_0. In this space, the periods of all pendulums, of all lengths, and on all planets collapse onto a single universal curve. Hence, dimensional analysis shows that all are governed by the same underlying physics. We can imagine using such experimental data to "close" our dimensional analysis and obtain an empirical fit to experimental data in the space of gt_0^2/s and θ_0.

Note how the experimental data (and the accurate prediction) in the bottom plot of figure 5.5 approach a constant value as the initial angle θ_0 becomes small. The near-zero slope of the curve in this region (small values of the abscissa) shows that the nondimensional parameter gt_0^2/s is insensitive to the values of θ_0 (see section 2.4.1). Hence, the experiments strongly suggest an asymptotic approximation of gt_0^2/s. Given the experimental evidence, we can be confident of an asymptotic limit for small angles of the form

$$\frac{gt_0^2}{s} = F(\theta_0) \approx \text{constant} = c_1 \text{ (for small } \theta_0).$$

Given the experiment data, the approximate value of c_1 is value of the constant is about

$c_1 \cong 39$ (extrapolation from experiments).

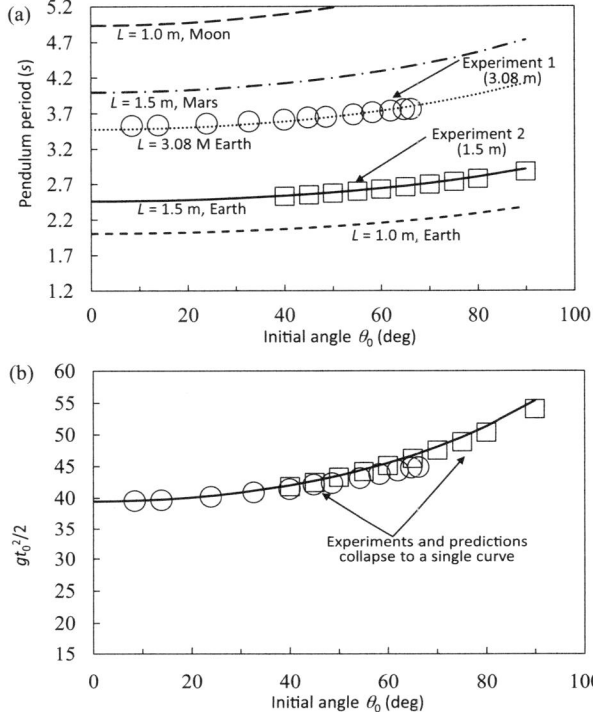

Figure 5.5
Collapse of pendulum period data (across initial angles, experimental methods, and planets). The top plot shows dimensional periods (in seconds) versus initial angle for pendulums of various lengths L. The circles and squares are experimental data from respectively Fulcher and Davis (1976) and Lima and Arun (2006), and the curves are predictions based on an approximate solution of the large-angle pendulum dynamics problem. The bottom plot shows the same experimental data and predictions but plotted as the nondimensional parameter gt_0^2/s versus initial angle θ_0. The prediction curves overlap exactly and appear as a solid line. More importantly, the experimental data collapses (to within experimental error); highlighting the universal nature of the new variables gt_0^2/s and θ_0.

What is the exact value? Well the exact solution for the period of a small angle displacement, point-mass pendulum with zero friction and drag is

$$t_0 = 2\pi\sqrt{\frac{s}{g}} \text{ (exact solution, small } \theta_0 \text{)}.$$

Solving for the nondimensional parameter, we find an irrational value

$$\frac{t_0^2 g}{s} = (2\pi)^2 = 39.47\ldots$$

Last, note that another application of the solution for pendulums of small displacements is as a measure of the Earth's eccentricity. The gravitational acceleration varies at different locations on the Earth. So, for example, taking the same pendulum to two locations a and b on the Earth, we have

$$\frac{\mathcal{E}_a}{\mathcal{E}_b} = \frac{c_1 s \, t_b^2}{c_1 s \, t_a^2} = \left(\frac{t_b}{t_a}\right)^2 .$$

Isaac Newton used this approach to quantify Earth's sphericity and discusses this method in the *Principia* (1687 [1850]). Newton obtained and analyzed experimental data (including his own) and wrote,

> Wherefore since the lengths of pendulums vibrating in equal times are as the forces of gravity, and in the latitude of Paris, the length of a pendulum vibrating seconds is. ... Now several astronomers, sent into remote countries to make astronomical observations, have found that pendulum clocks do accordingly move slower near the equator than in our climates.

5.9 Summary

- We began this chapter with a set of proposed rules D1–D5 applicable to the selection and, importantly, the exclusion of (dimensional) variables with which to describe a physical process. Identification of variables is the most difficult step of dimensional analysis and takes the most experience. It is essential to identify a sufficiently long list of variables while carefully also avoiding overspecification of the problem.

- In chapter 6, we will propose complementary rules ND1–ND6 for the simplification and analysis of functions of nondimensional parameters.

- Ipsen's method is an intuitive process that can be used to identify the nondimensional parameters of a problem and can be summarized as follows:

  1. Identify the n key variables associated with the problem.

  2. Identify p primary dimensions (e.g., M, L, T).

  3. Hypothesize a function of interest as a dependent variable and $n-1$ independent variables. Express primary dimensions for each.

  4. $n-p$ is a good guess for the number of nondimensional parameters.

  5. Choose a primary dimension to eliminate.

  6. Choose a variable (or group of variables) containing this primary dimension.

  7. Multiply or divide the chosen variable (to the correct power) by each variable in the current function as needed to cancel the chosen primary dimension. Do nothing with variables that do not contain the dimension chosen in Step 5.

8. Go to step 6, choose a new dimension, continue until you eliminate all primary dimensions.

- To illustrate both the power and the limitations of dimensional analysis, we explored a set of contrived problems (a falling stone, planet motion, and a pendulum) wherein we know analytical solutions. We contrived problems where we neglected these solutions to see how far dimensional analysis could take us.

- In each problem in this chapter (and in this book), we emphasized the importance of combining *dimensional analysis* with *physical intuition* and *experimental observations* to reduce the complexity of the problem.

- The falling rock and pendulum example analyses each showed how combining dimensional analysis with experiments is a powerful combination that can be leveraged to achieve the following: 1. Completely close (i.e., solve) a problem including determination of the form of the hypothesized function and associated constants; 2. Use dimensional analysis to collapse of experimental data and thereby use it how dimensional analysis to reduce the number of dimensions in a problem; and 3. Identify the key physics of a problem in such a way that the physics are comparable across experimental setups, experimentalists, absolute scale, thermophysical properties, and so forth.

- The example of Kepler's third law helped emphasized that we should include universal constants with dimensions in our hypothesized function, and it demonstrated experimental data collapse for observable planetary time scales.

Problems

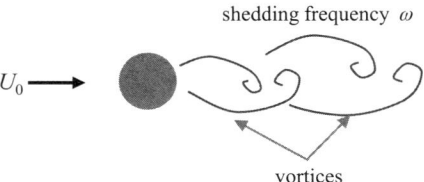

5.1.   A vortex can be described as some coherent region of fluid that spins about an axis in some frame of reference. Vortices and vortex shedding phenomena in fluid flow are ubiquitous and vary widely in size and shape, from tiny vortices shed by swimming goldfish to the Great Red Spot, a hurricane-like storm 20,000 km in diameter swirling around in Jupiter's atmosphere. As with a spinning top, the rotational inertia of such vortices persists until it is dissipated by friction (the effect of viscosity) to the surrounding air. Consider a smooth cylinder in a cross flow as shown in the following schematic. The approach flow is a uniform fluid velocity U_0. At sufficiently high Reynolds numbers, the cylinder interacts with the fluid flow and sheds vortices at some dominant frequency ω.

Apply Ipsen's method to derive a relation for the dominant vortex shedding frequency ω as a function of relevant variables in the problem. Since this may be your first experience with Ipsen's method, approach the problem in a systematic way as follows. First, fill in the blanks in the following table, which identifies the critical variables in the problem:

Variable	Definition	Physical quantity
ω	frequency of vortex shedding	T^{-1}
U_0	approach flow velocity magnitude	LT^{-1}
—	density of fluid	—
—	dynamic viscosity of fluid	—
—	diameter of cylinder	—

You may want to check with section 3.3 for the definition and dimensions of these variables. Next, hypothesize the unknown function in the following format:

$\omega = f(U_0, \_\_, \_\_, \_\_).$

Perform the step-by-step procedure to derive a nondimensional version of this function. When done, you may need to apply concepts we will present in section 6.1 to cast your derived function in terms of a nondimensional Strouhal number: $St = (\omega D/U_0)$ is a Strouhal number (a nondimensional frequency) and $Re = (\rho U_0 D/\mu)$. See Fey (1998) for a large amount of experimental data tabulated in terms of these parameters, and the impressive collapse of experimental vortex-shedding frequency data to a single, universal curve applicable across absolute length and velocity scales, fluid types, and associated thermophysical property data.

5.2. The critical variables describing the idealized (perpetual!) oscillation of a linear spring (spring constant k) and mass system on a frictionless surface are identified in this sketch:

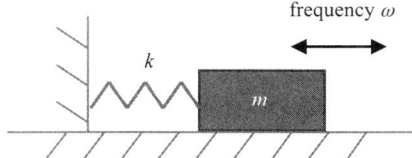

Use Ipsen's method to analyze the characteristic frequency of oscillation. How many nondimensional parameters are involved?

5.3.   The Earth is an oblate spheroid. Hirt et al. (2013) presented high-resolution esti-
        mates of the variation of the gravitational acceleration constant along Earth's sur-
        face. The minimum value was for the Nevado Huascarán summit in Peru, reported
        to be 9.764 m²/s; and the maximum value for the surface of the Artic Sea, reported
        to be 9.834 m²/s. Given these extremes, estimate the accuracy of the timing (in sec-
        onds) required by Isaac Newton to be able to estimate variations in Earth's radii
        from pendulum period measurements.

6

Combining Dimensional Analysis with Physical Intuition and Experimental Observations

Suppose the uniform force of gravity to tend directly to the plane of the horizon, and the resistance to be as the density of the medium and the square of the velocity conjunctly.

—Isaac Newton, *Principia* (1687)

This chapter begins with further formalization of the process of combining dimensional analysis with physical intuition and experimental observations. In chapter 5, I proposed rules D1–D5 around the process of selecting (and excluding) the list of dimensional variables that describe physical phenomena. In this chapter, I propose a set of six additional rules, rules ND1–ND6, which are useful in applying physical intuition and experimental observations to the task of simplifying the primary result of dimensional analysis: a hypothesized function of relevant nondimensional parameters.

After presenting rules we will analyze illustrative and, in some cases, famous examples of the application of dimensional analysis and its combination with physical intuition and experimental observations.

6.1 Rules of Thumb for Manipulating Functions of Nondimensional Variables

I will begin with a set of six proposed rules of thumb applicable to functions of nondimensional variables. These rules are used and referenced throughout the book as we combine dimensional analysis with physical insight. I term these rules ND1, ND2, and so forth to denote that they apply primarily to the manipulation and simplification of functions of nondimensional variables.

As mentioned in chapter 5, these proposed rules are my effort to formalize the process by which dimensional analysis can be combined with physical intuition and experimental observations to simplify and even solve problems. Note how each assumes some knowledge about or intuition regarding the physics of the problem and then provides a specific technique by which this intuition can be applied to the analysis.

6.1.1   Rule ND1: Reorganize Expressions of Nondimensional Variables

Recall from section 2.4 the vast generality and arbitrary nature of expressing an unknown function such as

$$Y = F(X).$$

This same generality helps us manipulate this expression and posit new functions that relate the relevant variables. We shall see that this is particularly useful for functions of non dimensional variables. For examples, we may want for convenience to express a related func tion between the square of Y and the inverse of X. Reasons for such manipulations include

- To ensure that the analysis ends up with traditional/recognizable nondimensional parameters such as Reynolds number, drag coefficient, Froude number, and so on.

- To isolate a variable such that it appears in one or more nondimensional parameters and not others

- To isolate known or weak dependencies

- To invert a function of a constant

To manipulate our nondimensional parameters, we recognize that each nondimensional quantity can be multiplied by itself or by another raised to a nonzero power, and the result is also nondimensional. Consider some function relating the nondimensional dependent variable H to nondimensional parameters X, Y, and Z:

$$H = F_1(X, Y, Z).$$

H^a, X^b, Y^c, and Z^d are each nondimensional for nonzero real values of a, b, c, and d. Similarly, the product $H^a X^b Y^c Z^d$ is also nondimensional for such exponents. Hence, we can posit any of the following

$$H^a = F_2(XY^b, Y, Z),$$

or

$$H = F_3(X, YZ^b, Z),$$

or

$$HX^a = F_4(X, YZ, X^a Y^b Z^c),$$

as three of many examples and for nonzero values of the exponents. Again, a, b, c, and d are arbitrary (real) and nonzero powers of the nondimensional parameter (and may be posi tive or negative).

In applying this principle, note that we must respect and preserve the number of dimensions of the function. Hence, we cannot simply eliminate a nondimensional quantity by dividing by itself (i.e., raising it to the 0th power).

Last, dimensional analysis sometimes leads to an expression as follows:

$$1 = F(X),$$

where a function of a nondimensional variable X is equal to unity. We can reorganize this function by inverting it as follows

$$X = \text{constant.}$$

That is, if a function of a variable is always equal to a constant (here unity), then the variable must itself be some constant.

6.1.2   Rule ND2: Isolate, Then Evaluate, Known Dependence

This rule is not easily used, but it is very powerful when applicable. If the form of the function is known for one of the independent variables x (holding the others constant), then this dependence may be evaluated explicitly. This evaluation is most successful if the function of nondimensional variables is manipulated to have only a single nondimensional term X, which contains the variable in question x. The known dependence may arise from the following:

- Experiments where the independent variable is varied while others are held constant
- A strong analogy with a similar, known problem is invoked
- Hypotheses are posed due to physical insights into the problem

Perhaps the most commonly hypothesized "known dependence" for an independent variable x is a power law of the form x^a.

In terms of our generalized notation, for some function of dimensional parameters:

$$r = f(a, b, c, d, e).$$

Consider a hypothetical case such that r is known to be directly proportional to b raised to the sth power as in

$$r \propto b^s,$$

in the case where the other variables are held constant. If the corresponding function of the nondimensional variables is, for example,

$$R = F\left(A = \frac{a}{b}, B = \frac{ba}{c^2}, C = \frac{d}{e} \right),$$

then we have not isolated the variable in question (b) and we cannot use this rule. If we manipulate the function (as per rule ND1) to derive

$$R = G\left(\frac{a}{b}, \frac{a^2}{c^2}, \frac{d}{e} \right),$$

then the dependence on b is confined to the first term within the parenthesis. If the other variables b, c, d, and e do not depend on b (e.g., as per rule D2 in section 5.1.2), then we can hypothesize that

$$R = \left(\frac{a}{b}\right)^{-s} H\left(\frac{a^2}{c^2}, \frac{d}{e}\right).$$

That is, we explicitly and formally write the mathematical (directly proportional) dependence for the sole nondimensional parameter that contains the dimensional variable of known dependence. This dependence is formulated as a prefactor multiplying the unknown function. Note that this respects the principle of dimensional homogeneity, since both the dependent variable (on the left-hand side) and the prefactor are dimensionless. Also note that the key is to isolate the quantity of known dependence (here b). Given this, the non-dimensional parameter that contains this parameter of interest can contain other quantities (here a).

The following is an example I have adapted from Rayleigh's (1915) brief paper on dimensional analysis. Consider a function describing the total heat transfer (energy per time) from a long thin wire immersed in a fluid with specific heat c_p, density ρ, conductivity k, moving at velocity V. For this problem, Rayleigh leverages what I have termed rule D1 and begins his analysis by hypothesizing the following function

$$\dot{Q} = f(a, L, \Delta T = T - T_\infty, V, \rho c_p, k).$$

Here L and a are the length and radius of the wire, respectively. As per rule D1, Rayleigh has already intuited that the problem will not depend on two temperatures individually, or on density and specific heat individually, but in terms of a temperature difference $T - T_\infty$ and the product ρc_p, respectively. Rayleigh's use of our rule D1 has already reduced a nine-variable problem to a seven-variable problem. A nondimensional analysis (e.g., using Ipsen's method) for such a hypothesized function yields:

$$\frac{\dot{Q}}{kL\Delta T} = F\left(\frac{aV\rho c_p}{k}, \frac{L}{a}\right).$$

Next, Rayleigh pursues the hypothesis that the dimensional heat transfer rate $\dot{Q}$ will scale proportionally as some power of velocity V. He uses this information and what we here term rule ND2 to simplify the expression. Rayleigh attributes this physical insight to the theoretical work of J. V. Boussinesq (1905), who presented an approximate closed-form analytical solution to this problem for the case of very thin thermal boundary layers (and inviscid flow). In such a limit, the situation is nearly one-dimensional heat transfer. For example, Boussinesq solves the problem by approximating a streamline of inviscid flow around the cylinder as being equivalent to inviscid flow over a flat, finite wall with a length equal to half the cylinder's circumference. In that simpler case, the problem is that of an energy transport boundary layer subject to a uniform and steady-slip flow) and the heat transfer rate scales as follows with velocity

$$\dot{Q} \propto V^{1/2}.$$

To apply rule ND2, we first make sure that the dependent variable in question is confined to a single nondimensional parameter. In this example, velocity appears only within the first nondimensional parameter, $aV\rho c_p/k$, so this condition is satisfied. Second, we note that the rest of the variables in this parameter (a, c_p, and k) do not depend on the identified variable V. Third, we infer that the dependence will apply to the entire nondimensional term involving velocity. Fourth, we explicitly evaluate the function as

$$\frac{\dot{Q}}{kL\Delta T} = \left(\frac{a\rho V c_p}{k}\right)^{1/2} G\left(\frac{L}{a}\right).$$

This idea can be immensely useful in simplifying functions of nondimensional variables. We should stress that this partial evaluation of the function will only hold if the dimensional dependence holds (here $\dot{Q} \propto V^{1/2}$). Of course, the true test of the approach is to see if the scaling collapses *experimental data*.

A second example of this principle will be discussed in chapter 8 in the context of pressure drops in pipe flow where the known (directly proportional) dependence on pipe length is evaluated explicitly.

6.1.3   Rule ND3: Isolate a Variable with an Unknown but Weak Dependence

If the function is a very weak function of one parameter, then work to have that parameter appear as little times as possible. Consider again the drag of a sphere of diameter d traveling at velocity V. For high speeds, we might hypothesize the drag force is given by

$$F = f(d, V, \rho, \mu, c),$$

where ρ, μ, and c are respectively the fluid density, dynamic viscosity, and the local speed of sound (a measure of the compressibility of the flow). A naïve application of dimensional analysis might yield the following

$$\frac{F}{\rho c^2 b^2} = F_1\left(\frac{\rho c b}{\mu}, \frac{V}{c}\right),$$

where V/c is a Mach number Ma. Although dimensionless, the variables are all "corrupted" by the speed of sound and we have lost our standard definitions of drag coefficient and Reynolds number. Further, if we believe the phenomena will be a weak function of Mach number (e.g., Ma less than roughly 0.3), we then want to isolate the weak functionality. To remedy this, we invoke rule ND1 in section 6.1.1 and manipulate the variables as follows:

$$\frac{F}{\rho c^2 b^2}\left(\frac{V}{c}\right)^{-2} = F_2\left(\frac{\rho b c}{\mu}\cdot\frac{V}{c},\frac{V}{c}\right),$$

so

$$\frac{F}{\rho V^2 b^2} = F_3\left(\frac{\rho b V}{\mu},\frac{V}{c}\right).$$

The possibly weak dependence on speed of sound is now confined to the third parameter.

In some limited cases, we may be able to eliminate variables in some limit, but only if we have some knowledge that the function has a weak dependence of said variable in this limit. Here, in the limit of small Ma, knowledge of the physics suggests that drag indeed becomes insensitive to Mach number, and so we may write

$$\frac{F}{\rho V^2 b^2} \simeq F_3\left(\frac{\rho b V}{\mu}\right).$$

Such elimination should be done with great care and requires knowledge of the dependence (see rule ND6), or detailed experimental observations (e.g., where Ma is varied while Reynolds number is held constant, and vice versa).

6.1.4 Rule ND4: Replace a Nested Function with Its Independent Nondimensional Parameters

After performing dimensional analysis, you may subsequently convince yourself that one of the nondimensional parameters can be expressed as a function of the others. For example, this may become apparent only after you performed dimensional analysis. Also, you may have new experimental evidence confirming such dependence. Consider the following function of nondimensional parameters A, B, C, and D:

$A = F(B, C, D).$

If you can define some known or unknown function such that $B = G(C, D)$ (or confirm such experimentally), then you can simplify the hypothesized function to

$A = G(C, D).$

6.1.5   Rule ND5: Consider Absorbing an Approximately Constant Nondimensional Variable into Function

In general, as we mentioned in rule D3, we may absorb nondimensional parameters that are exactly constant into a function (e.g., π) without loss of generality. This is always appropriate. The constant now just becomes part of the definition of the function. For example, for geometric similarity, we implicitly absorb all of the dimensionless and definable length, area, and volume ratios of the geometry into the function.

In some cases, you may consider applying this approach in an approximate fashion. Hence, we can simplify a function of nondimensional variables by absorbing one whose variations have only negligible effect on the dependent variable. In this way, we treat the nondimensional variable as approximately constant. Consider the function

$$A = F(B, C, D).$$

If variations of the nondimensional parameter D (e.g., over the full range of expected values of D), have negligible effect on A (compared to the effect of B and C), then consider treating this nondimensional variable as if it is a constant and absorbing it into the function and the associated hypothesis that

$$A \cong F(B, C).$$

Again, such hypotheses may arise from experimental evidence, analogies with other problems, or physical intuition about the problem.

6.1.6   Rule ND6: Be Careful Eliminating Variables—Even If They Are Small

It is not appropriate to eliminate a variable just because it is small. Such elimination is only possible with some knowledge of the function or with some physical insight. Consider the following hypothetical function in terms of nondimensional variables A, B, C, and D:

$$A = F(B, C, D).$$

Consider a limit such that D approaches small values. This limit alone is not sufficient to determine the strength or weakness of the function's dependence. For example, consider that the underlying function may have the form

$$A = CD + BD.$$

Hence, the effect of D on A is strong in the limit of vanishing D. Also, if the limiting behavior also affects the magnitude of C or B, then we must evaluate the product CD versus the product BD in such a limit. Knowledge of the physics is required to eliminate variables.

6.2   Spine Patterns of Liquid Drop Impacts and Blood Drop Patterns in Forensics

We continue our discussion of dimensional analysis with an example problem that illustrates its utility and demonstrates application of some of the eleven rules proposed so far in chapters 5 and 6. Consider the splatter patterns of liquid drops moving through air and impacting on a hard surface. This splatter results in "spines" that radiate roughly from the center of the drop. Figure 6.1a shows images from experiments of this phenomena. These experiments were performed by me using a single-channel (air-displacement type) pipette to create 10 microliter droplets (diameter $d = 2.7$ mm) from heights of 0.15, 0.30, 0.64, 1.24, 1.55, and 1.85 m. The drops impacted on sheets of construction paper and were imaged using a smart phone camera.[1] These shapes can be visualized fairly easily at home by dropping water or milk on a piece of paper.

The droplet spline phenomena is common, from the first few raindrops hitting the sidewalk, to drops of sweat hitting a gym floor. The shape of such drops are also relevant to forensic studies. For example, blood drop shape can provide evidence as to the velocity and/or trajectory of blood splatters at a crime scene. Adam (2012) and Attinger et al. (2013) present reviews of the fluid dynamics associated with blood stains and blood splatter in forensics, although the analysis presented here applies to all liquids and not just blood.

We begin the analysis with a physical insight.

Physical insight: The drops are due to an instability in the fluid dynamics associated primarily with a competition between inertial forces and surface tension forces. Hence, we can expect that the pattern will be insensitive to liquid viscosity.

Hence, we base our function for the number of spines in terms of a hypothesis that the following variables are essential and sufficient:

Variable	Definition	Physical quantity
N	number of spines (dependent variable)	—
V	velocity of drop upon impact	LT^{-1}
ρ	density of liquid	ML^3
σ	surface tension	MT^{-2}
d	diameter of drop (before impact)	L

1. Unpublished experimental visualization of droplet splatter patterns on the surfaces of sheets of red construction paper. Droplets were created using a 1–100 μL single-channel pipette and heights were measured with an ordinary tape measure. Impact velocity was estimated from the drop heights and is listed in the main text neglecting the effects of drag for these short drops. Water density and surface tension were taken as 998 kg/m³ and 7.29 N/m, respectively. A Samsung Galaxy S8 phone (model SM-G950U) was used to obtain the images, and each image was obtained within about 2 s after drop impact.

Note how geometric similarity is here applied in an approximate sense only (i.e., the drops are may be geometrically similar immediately prior to impact, but the splatter patterns are not strictly similar). We hypothesize the function

$$N = f(V, \rho, \sigma, d).$$

Note that N is dimensionless but certainly not a constant, so we cannot apply rule D3. We can use dimensional analysis to show this can be written as

$$N = F\left(\frac{\rho V^2 d}{\sigma}\right).$$

Here we find a so-called Weber number relating inertia and surface tension as

$$We = \text{Weber number} = \frac{\rho V^2 d}{\sigma} = \frac{\rho V^2 d^2}{\sigma d} = \frac{\text{inertial forces}}{\text{surface tension forces}}.$$

Inertial forces enhance fingering and therefore spines, while surface tension reduces fingering. The preceding relation and the associated interpretation of the Weber number are as far as dimensional analysis alone can take us. We next consider insights based on experimental observations analyzed in the context of the approximate dimensional analysis (e.g., which only approximated geometric similarity and which neglected the effect of viscosity).

Figure 6.1a shows experimental images of water drops on a surface of construction paper. The scale bar in the first image applies to all other images. Shown together with each image is the value of the square root of Weber number, $We^{0.5}$, for each experiment. As in our analysis, the Weber number is defined based droplet diameter and the estimated velocity upon impact (see footnote to section 6.2). Figure 6.1b shows a plot of the experimental data of Adam (2012). Plotted is Adam's measured number of spines as a function of the square root of Weber number. This square root dependence can be discerned by, for example, plotting N versus the log of Weber number and noting the slope. The power law $We^{0.5}$ well collapses the experimental data into a tight line for We below about 900 ($We^{0.5} = 30$). At higher We, the data deviate from the power law and exhibit more scatter.

Next, we will consider the possible effects of viscosity. Note the significantly greater vertical scatter in the data for high We. That is, there is apparently a range of values of N for each value of We. As discussed in section 4.3, such observation of "multiple experimental values" of the dependent variable (N) for a fixed We is a fairly good indication that the dimensional analysis has, in this regime of high We, neglected an important variable. A hypothesis for such deviation (which is consistent with the observations of Attinger et al., 2013) is that the experiments associated with these high We are somehow dependent on the effects of viscosity, which was ignored in the analysis. At the high values of We, the simple two-variable solution $N = F(\rho V^2 d/\sigma)$ is plainly insufficient to collapse the data.

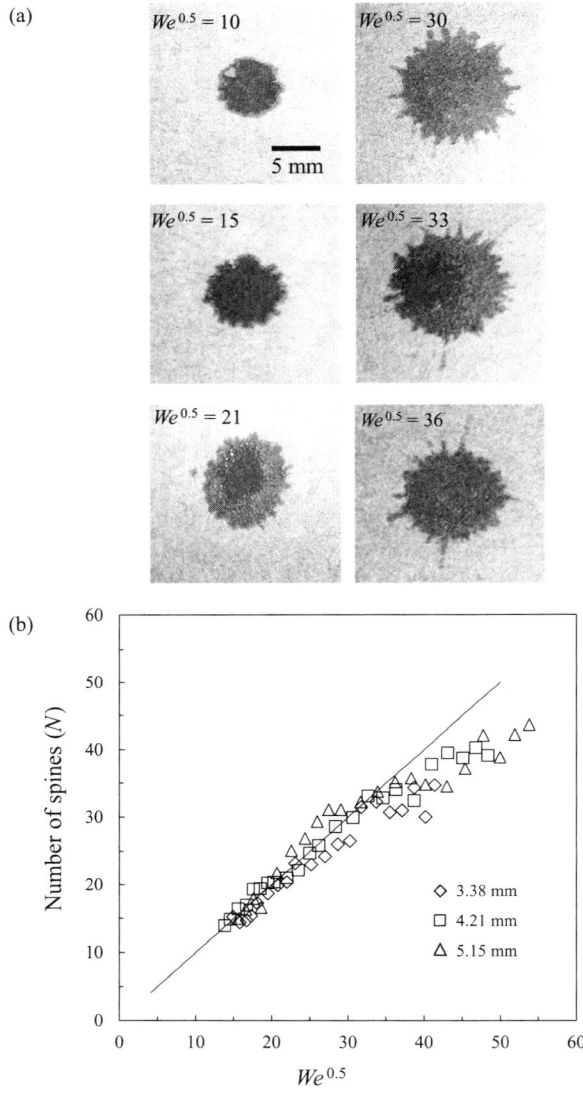

Figure 6.1
Fluid dynamic instabilities during splatter (onto a surface) of a liquid droplet lead to so-called spines that radiate from the drop (left). Dimensional analysis leads to a correlation between the number of spines and Weber number. The data collapses very well to a straight line in terms of square root of Weber number for low We, but deviates from this line at higher values (likely due to the effects of viscosity). The straight line shown is a scaling directly proportional to the square root of We (linear regression between the origin and $We^{0.5} = 30$). Images were obtained by me and described in the footnote to section 6.2. The data of (b) were here digitized from Adam (2012). The values in the legend refers to droplet diameter (prior to impact).

We might then revise our theory for this regime as follows. We form a new list of essential variables as follows

Variable	Definition	Physical quantity
N	number of spines (dependent variable)	—
V	velocity of drop upon impact	LT^{-1}
ρ	density of liquid	ML^3
σ	surface tension	MT^{-2}
d	diameter of drop (before impact)	L
μ	dynamic viscosity	$ML^{-1}T^{-1}$

Accordingly, we posit the following function

$$N = f(V, \rho, \sigma, d, \mu).$$

We then apply Ipsen's method (left as an exercise to the reader) to derive

$$N = F_2\left(\frac{\rho V^2 d}{\sigma}, \frac{\mu}{\rho V d}\right).$$

Then, we can apply rule ND2 to derive a more conventional form:

$$N = G\left(\frac{\rho V^2 d}{\sigma}, \frac{\rho V d}{\mu}\right),$$

where $\rho V d/\mu$ is a Reynolds number, Re. We see that including viscosity has increased the complexity of the hypothesized function of nondimensional parameters from two variables to three. Presumably, we can then perform a new set of experiments and see if we can systematically collapse single values of N to a surface defined by We and Re. To my knowledge, such systematic experimental study of this problem has not been published.

Last, note that images of higher quality than those obtained by me in Figure 6.1a are presented in the review by Attinger et al. (2013) (see figure 9 of that reference). Attinger et al. (2013) presents high-speed sequential images of drops as they impact a surface. An interesting challenge associated with such experiments is that spines of higher Weber number drop impacts initially project outward (in a 2 ms time period) but then quickly retract back within about 20 ms after impact. This rapid retraction is due to the effects of surface tension and results in high velocity and fast dynamics. Hence, precise and quantitative high-speed imaging of the process is required for highest-quality, quantitative experimental data.

6.3   Atomic Explosions: An Example Confirmation of a Power Law

6.3.1   History of Atomic Explosion Analysis by G. I. Taylor

A famous example application of dimensional analysis is around an analysis by British physicist (and fluid mechanician) Geoffrey Ingram Taylor. The details of this story are discussed in detail by Deakin (2011) and only summarized here. In 1941, Taylor was approached by the UK Ministry of Home Security to speculate on the explosive yield of the USA's newly demonstrated nuclear fission bomb. Taylor did not begin with a dimensional analysis but instead created a model based on inviscid Navier–Stokes equations in spherical coordinates (including continuity), an equation of state, and the Rankine–Hugoniot jump conditions for the shock wave. Taylor's equations related energy of the detonation to its radius versus time and assumed an initial "point source" (energy released within a very small space and within a very short duration) and a spherical detonation wave. In the early 1940s, this problem was also studied by John von Neumann in the United States and Leonid Sedov in the Soviet Union. The latter two theoretical analyses were in a way more successful than that of Taylor in that they found exact solutions of the respective approximate mathematical models given similar assumptions.

Taylor's solution, however, is described here because of his validation of the scaling law by using experimental data; an essential aspect of dimensional analysis. Comparison with experiments and the associated collapse of experimental data in nondimensional space is the ultimate validation of a scaling law.

As part of his analytical formulation, Taylor found an approximate solution of the form

$$R = \left(\frac{Et^2}{\rho K} \right)^{1/5} . \tag{6.1}$$

This can be inverted to solve for the explosive yield energy E to yield

$$E = K \frac{\rho R^5}{t^2} . \tag{6.2}$$

Here the radius of the blast R should initially depend on only E, the time t after the detonation, and the density ρ of the surrounding air. K was an unknown parameter that Taylor estimated would be a function of the specific heat ratio of the gas within the cloud. Deakin points out that this relation is found explicitly in Taylor's report but only implicitly in those of Neumann and Sedov. Taylor submitted his report to his government in June 1941, and he was not cleared to publish this work until 1949.

The story continues in 1947 when the US Atomic Energy Commission released a film of the atomic bomb Trinity test, along with 25 still photographs (Mack, 1947). The photos showed the detonation wave at various times and associated shape of the detonation wave (see examples in figures 6.2 and 6.3). Such photos also appeared in *Life* magazine and other popular publications.

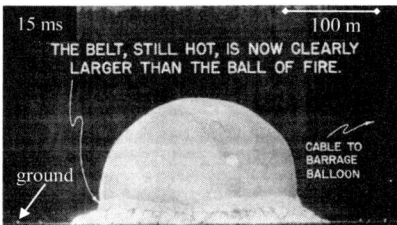

Figure 6.2
Images of trinity atomic bomb explosion of 1945 published by the US Atomic Energy Commission (Mack, 1947). Note the time stamps, scale bars, and location of ground level. The text regarding the cloud's "belt" and the cable are from the original Mack reference. G. I. Taylor used information from such published images to estimate the explosive yield of the weapon.

Taylor used these published data to analyze the dynamics of the fireball. To do this, he took the logarithm of both sides of equation (6.2) to derive (Taylor, 1950)

$$\frac{5}{2}\log_{10} R = \log_{10} t + \frac{1}{2}\log_{10}\frac{E}{\rho K}. \tag{6.3}$$

Since the last term is approximately a constant (for any given detonation under the assumptions), a plot of $\log_{10} R$ vs. vs. $\frac{2}{5}\log_{10} t$ should yield a straight line. Taylor showed this relation was a very good fit to the experimental observations, as shown in figure 6.3. This collapse of data validates the scaling law predicted by the theory.

The comparison to experimental data provided Taylor a close estimate for the extrapolated value of y-intercept of the fit; that is, a value of

$$\frac{5}{2}\log_{10}\frac{E}{\rho K} \approx 7.9,$$

from which he estimated the following implicit relation

$$E = (8.45E13)K.$$

Deakin discusses how Taylor, Sedov, and Neumann each estimated from knowledge of the physics of shock waves that the value of K would depend on the ratios of specific heats, γ, of the fluid within the blast. In fact, each independently used analytical models to estimate the functional relationship between K and γ. The three researchers independently derived estimates for K within the narrow range of 0.851 to 0.856. It is important to note that dimensional analysis alone did not and could not lead to an estimate of this constant. Neumann's estimate of 0.8510 is believed to be the most accurate, and using it, we can estimate

$$E \approx 7.19\times10^{13} \text{ J} = 17 \text{ kilotons of TNT}.$$

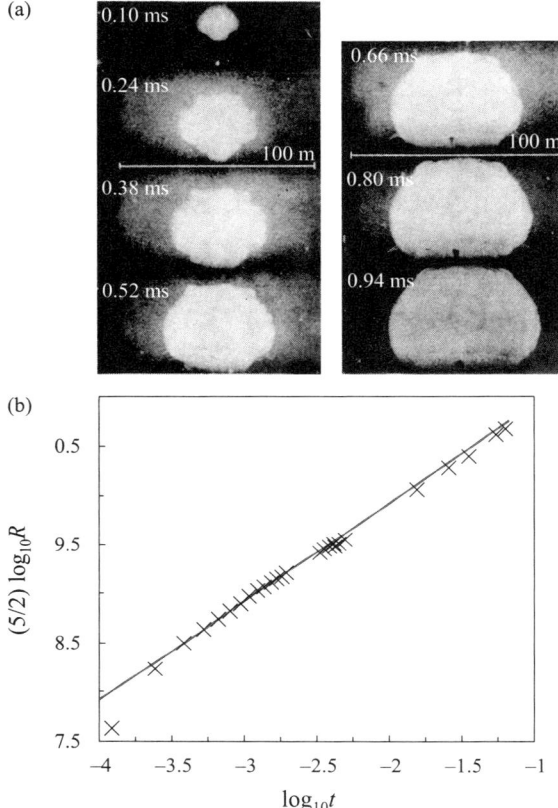

Figure 6.3
G. I. Taylor's data comparing blast radius and time. (a) Images from Mack (1946) of the first ~1 s of the Trinity explosion. The comparison between such data and the current scaling result from dimensional analysis are shown in (b). The power law fit has the form $\frac{5}{2}\log_{10} R$ vs. $\log_{10} t$. The data was digitized from Taylor (1950b). The y-intercept is 7.9.

6.3.2 Atomic Explosion Analysis Using Dimensional Analysis

The application of dimensional analysis requires two basic assumptions used by Taylor. First, the detonation wave is assumed to be spherical. Second, the energy is released within a relatively small space. The analysis further assumes that heat transfer (e.g., radiative energy loss) is not important in determining approximate explosive yield from radius data.

We begin by postulating the following essential variables:

Variable	Definition	Physical quantity
E	Energy released from a point	ML^2T^{-2}
R	Radius of detonation wave	L
t	Time since detonation	T
ρ	Density of air	ML^3

We hypothesize the desired unknown function as:

$R = f(E, t, \rho)$.

This results in a single nondimensional parameter, and so a function of the form

$$1 = F\left(\frac{\rho R^5}{Et^2}\right).$$

This leads to the hypothesis that the nondimensional group is a constant (see rule ND1) and that

$$E = c_0 \cdot \frac{\rho R^5}{t^2}.$$

It is important to note that this is as far as dimensional analysis alone can take us. It provides absolutely no way to estimate c_0, and we cannot estimate even the order of this constant.

As described earlier, it happens that the constant c_0 is in fact very close to unity and approximately equal to 0.85 (see Deakin for further discussion and note that our c_0 is Deakin's parameter K). For example, at $t = 6$ ms the radius of the shock wave is roughly 80 m. Substituting this and $\rho = 1.2$ kg/m^3 and a value of $c_0 = 0.85$ yields Taylor's estimate of 17 kilotons of TNT.

The actual yield of the Trinity test is not precisely known. Estimates of the yield of this bomb range from about 15 to 20 kilotons of TNT (Deakin, 2011). The estimate given here is within the modern, accepted (and public) range.

This example is most interesting because it highlights both the power and limitation of dimensional analysis. The hypothesized (and somewhat risky) short list of variables and dimensional analysis yields a function relation of the form

$$E = c_0 \cdot \frac{\rho R^5}{t^2}.$$

This relation demonstrably collapses the experimental data but dimensional analysis provides absolutely no way to estimate the value of this important constant.

6.4   World-class Weightlifters: Lifter-to-Lifter Comparisons and Lift Data Collapse

I here present a dimensional analysis applicable to weightlifting records. The idea for this analysis has been attributed to the late Joe B. Keller of Stanford University (as part of a seminar that the applied mathematician taught in 1973 at Stanford called "Mathematics Applied to Athletics"). However, I am not aware of Keller's analysis or result or even his general approach. Hence, presented here is my analysis of the problem.

Athletes who compete in Olympic and world-class weightlifting competitions are divided into weight categories. This approach is perhaps unnecessary given a correct scaling of each weightlifter's ability. The question here is: How do we scale the lifted weight so we can compare it to a lifter of another weight class?

Physical insight: The key insight here is that the weight you can lift scales as the product of a maximum muscle stress τ_m and a muscle cross-sectional area A. We will assume that lifters are approximately geometrically similar. We also assume that all of these athletes have the same mass density.

Given this insight, we proceed by applying rules D2 and D5. First, we assume geometric similarity across weightlifters. If so, then their volume (and weight) and cross-sectional area of any muscle can all be scaled with the same geometric parameter. Here we arbitrarily select the value of their height s. For example, given s and geometric similarity, weight scales as s^3. We therefore postulate the following variables

Variable	Definition	Physical quantity
m_l	lifted mass (dependent variable)	M
s	height of lifter	L
τ_m	maximum stress in muscle	$ML^{-1}T^{-2}$
g	gravitational acceleration	LT^{-2}
ρ	body density of lifters	ML^{-3}

As per rule D2, after including a length scale, we can include either the body mass of lifters or their density. We here arbitrarily chose the latter. As per rule D5, once we define geometric similarity and use these parameters, we should not also include the mass of lifters m_0. This is a crucial aspect of dimensional analysis: Including redundant variables leads to superfluous nondimensional parameters and less insight.

We can summarize the relevance of rules D2 and D5 as follows. A good check is to make sure that each variable not included follows directly from a variable included and geometric similarity. In the case of weightlifters, the mass of lifters is simply $m_0 = c_0 \rho s^3$, where c_0 is a nondimensional constant that follows from geometric similarity, so we exclude m_0.

We hypothesize a function of the form

$$m_l = f(s, \rho, g, \tau_m).$$

We perform dimensional analysis and derive

$$\frac{m_l}{\rho s^3} = F_1 \left(\frac{\tau_m s^2}{m_l g} \right).$$

Note that the length scale s appears in both terms. We can rewrite the dependent variable as follows:

$$\frac{m_l}{\rho s^3} = \frac{m_l}{m_0 / c_0},$$

where c_0 is the aforementioned nondimensional ratio accounting for the specific geometry of our geometrically similar lifters. As per rule ND5, we can absorb this constant into a new function as follows:

$$\frac{m_l}{m_0} = F_2 \left(\frac{\tau_m s^2}{m_l g} \right).$$

Further, we can use the definition of c_0 to eliminate the direct dependence on length scale as follows:

$$s^3 = \frac{m_0}{c_0 \rho} \implies s^2 = \left(\frac{m_0}{c_0 \rho} \right)^{2/3}.$$

Whence,

$$\frac{m_l}{m_0} = F_3 \left(\frac{\tau_m}{m_l g} \left(\frac{m_0}{c_0 \rho} \right)^{2/3} \right) = F_3 \left(\frac{\tau_m}{(c_0 \rho)^{2/3} g} \frac{m_0^{2/3}}{m_l} \right),$$

or, again using rule ND1, as

$$\frac{m_l}{m_0} = F_4 \left(\frac{(c_0 \rho)^{2/3} g}{\tau_m} \frac{m_l}{m_0^{2/3}} \right).$$

Here, we recognize that the group $\dfrac{\tau_m}{(c_0 \rho)^{2/3} g}$ as a sort of (dimensional) stress index that measures the quality of a lifter.

Figure 6.4 shows a plot from world record data obtained from the International Powerlifting Federation (IPF) as of 2018. Shown are lifted mass data for deadlift and bench press as a function of the mass of the lifter. The data are plotted both in their raw form and in the coordinates of $m_l / m_0^{2/3}$ vs. m_0, which are derived from the dimensional analysis. The

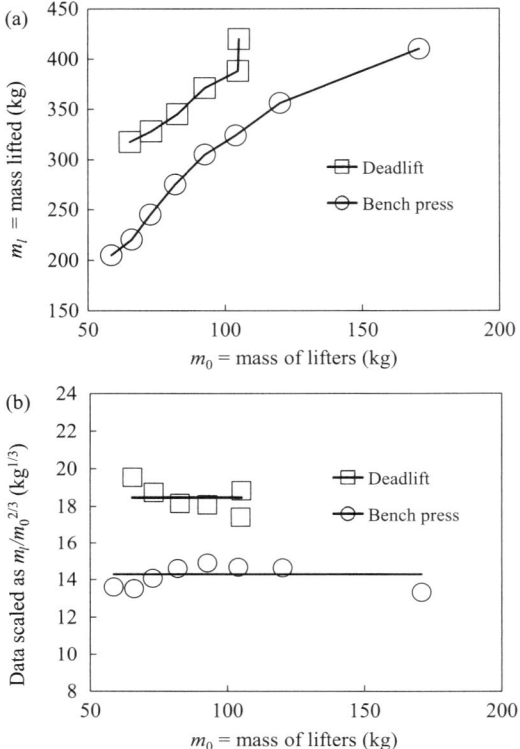

Figure 6.4
International Powerlifting Federation (IPF) world records for deadlift and bench press. Plot (a) shows raw data for mass lifted as a function of the mass of the lifter. Plot (b) shows the same data for lifted mass but as a function of $m_l/m_0^{2/3}$. Both plots show variations of a factor of 5 on the ordinance. The bottom plot includes horizontal lines at arithmetic averages of 18.5 and 14.3 for deadlift and bench press, respectively. The predicted scaling fairly well collapses the data, reducing the maximum-to-minimum variations from 1.3 to 1.1 for deadlift and from 2.0 to 1.1 for bench press.
Source: http://www.powerlifting-ipf.com (accessed July 2018).

scaling fairly well collapses the experimental data, particularly the bench press event. Note that both plots show a fivefold variation in the ordinance. The narrowing of the variations around the predicted scaling law suggests it may indeed be applicable across body weights.

6.5   Running (from) Dinosaurs: An Analysis That Does Not Assume Geometric Similarity

Can a time-traveling human outrun a dinosaur? To tackle this seemingly impossible question, we will first consider an analysis of the observable running speed of (nonextinct) animals. The data we will analyze is from M. R. Alexander (1991).

Physical insight 1:   Alexander (1991) presents a basic discussion of the dynamics of walking versus running, the two basic gaits of bipeds. For example, walking bipeds typically approximately fix the knee of the leg touching the ground, and so their motion is approximately a rotation about the foot contact. Think here of an upside-down pendulum with a pivot on the ground and cresting the top of its arc as shown in figure 6.5. L is the distance between the center of mass and the point of contact with the ground (approximately hip height). When walking, a vertical force balance of this simple model yields

$$N - mg = ma_y = -mV^2/L,$$

where N is the normal force exerted by the ground and $-mV^2/L$ is the centripetal acceleration. Since N is strictly positive (upward), the maximum speed for walking is then limited by $N=0$, where $mV^2/L = mg$. Therefore, the maximum tangential velocity of a walker's center of mass is limited to approximately

$$V^2/L \leq g \text{ (walking limit for biped)} \tag{6.4}$$

If the running speed is at the critical value of equation (6.4), then the force on the leg is zero (gravity force equal to centripetal force). Just above this speed, the incipient runner's foot leaves the ground and, at some point, neither foot touches the ground for some time—the difference between walking and running. Moving even faster requires greater times and distance where the body is not touching the ground. (Interestingly, the runner can walk just above this limit if he wiggles his hips like an Olympic walker so as to reduce the rise and fall displacements of his center of mass and partially avoid the pendulum-like dynamics.)

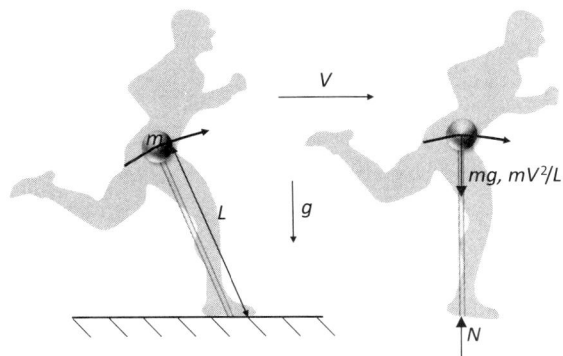

Figure 6.5
Schematic of a human just breaking into a run. The runner's mass m approximately pivots on the leg touching the ground. We roughly model the dynamics as those of an inverted, solid-rod pendulum. At the top of the arc, his maximum centripetal acceleration is the limited by his weight. A free-body diagram of the mass/rod system is superposed onto the sketch on the right.

The criterion of equation (6.4) can be expressed as $Fr \leq 1$, where $Fr \equiv V^2/(gL)$ is the nondimensional Froude number. Froude number can be interpreted as capturing the trade-off between kinetic and potential energies of motion. Consider

$$\frac{Fr}{2} = \frac{V^2}{2gL} = \frac{\frac{1}{2}mV^2}{gmL} = \frac{\text{kinetic energy}}{\text{potential energy}}. \tag{6.5}$$

Physical insight 2:   Although quadrupeds may be geometrically similar to each other, they are strikingly dissimilar to bipeds and have very different gaits from those of bipeds. As Alexander (1991) describes, quadrupeds have three basic gaits: walk, trot, and gallop. However, we here will hypothesize that the idea of a critical Froude number dividing walking versus running from bipeds may also be applied as criteria dividing walking versus trotting and trotting versus galloping in quadrupeds. If this is correct, then Froude number arises as a fairly general characterization of animal gait. The importance of Froude number will help us identify a small (perhaps minimum) number of relevant dimensional variables describing maximum running speed for all animals.

Here we first present a dimensional analysis of the maximum running speed of animals. We loosely follow Alexander (1983) and postulate the following parameters:

Variable	Definition	Physical quantity
U	velocity of a specific gait (e.g., running)	LT^{-1}
L_{stride}	Distance traveled in a single complete cycle of leg movements	L
L_{leg}	Hip height (distance from ground to hip)	L
g	gravitational acceleration	LT^{-2}

Note that Alexander's analysis is not based on geometric similarity. Quadrupeds and bipeds are dissimilar. Instead, the analysis is based on the basic idea presented earlier: That gait should be a strong function of Froude number (as per the inverted pendulum idea) and one other length scale (the stride length). Also, note that Alexander's choice of variables includes two independent length scales. The directions of these are orthogonal and leg length is a fixed geometric parameter for each animal, while stride length is determined with running speed. Alexander (1991) observes that hip height is typically nearly the same as height to the center of mass of a running animal. From these parameters, we hypothesize the following function of interest

$$U = f(L_{\text{stride}}, L_{\text{leg}}, g).$$

Note that we know we can exclude the mass of the pendulum, as per rule D4. We employ dimensional analysis to derive

$$\frac{U}{\sqrt{gL_{\text{leg}}}} = F_1\left(\frac{L_{\text{stride}}}{L_{\text{leg}}}\right).$$

Note that we derived a nondimensional parameter that is simply the square root of Froude number in terms of a leg length

$$\sqrt{\text{Fr}} = \frac{U}{\sqrt{gL_{\text{leg}}}}.$$

We can apply rule ND1 to invert this (single valued) unknown function as

$$\frac{L_{\text{stride}}}{L_{\text{leg}}} = F_2\left(\frac{U}{\sqrt{gL_{\text{leg}}}}\right). \tag{6.6}$$

Figure 6.6a shows a comparison between selected raw data for stride length versus velocity for three animals. The data is fairly linear for each case (a characteristic not determinable by dimensional analysis), albeit with different slopes and y-intercept values. Similar data are plotted in figure 6.6b in terms of the variables identified by dimensional analysis: $L_{\text{stride}}/L_{\text{leg}}$ versus square root of Froude number of the form $U/\sqrt{gL_{\text{leg}}}$. Shown together with the data is a linear regression (best fit) line. The data suggests that the function F_2 is fairly well approximated by a straight line (with finite y-intercept).

This fairly general collapse across species including bipeds and quadrupeds suggests that the scaling might also be applicable to dinosaurs. That is, to estimate the speed of dinosaurs, we hypothesize that the gaits of dinosaurs is expected to fall close to the regression line of figure 6.6b. We can obtain the slope m and y-intercept value b from a linear regression of the experimental data for nonextinct animals as follows:

$$\frac{L_{\text{stride}}}{L_{\text{leg}}} = F_2\left(\frac{U}{\sqrt{gL_{\text{leg}}}}\right) = m\frac{U}{\sqrt{gL_{\text{leg}}}} + b.$$

The linear regression is shown in figure 6.6 and suggests best-fit values of $b = 0.82$ and $m = 1.3$. Given this fit of the data for modern animals, we estimate the running speed of a dinosaur. From the fossil record of various dinosaurs, we know

- L_{leg} (e.g., fossilized leg bones)
- L_{stride} (e.g., tracks on tar)

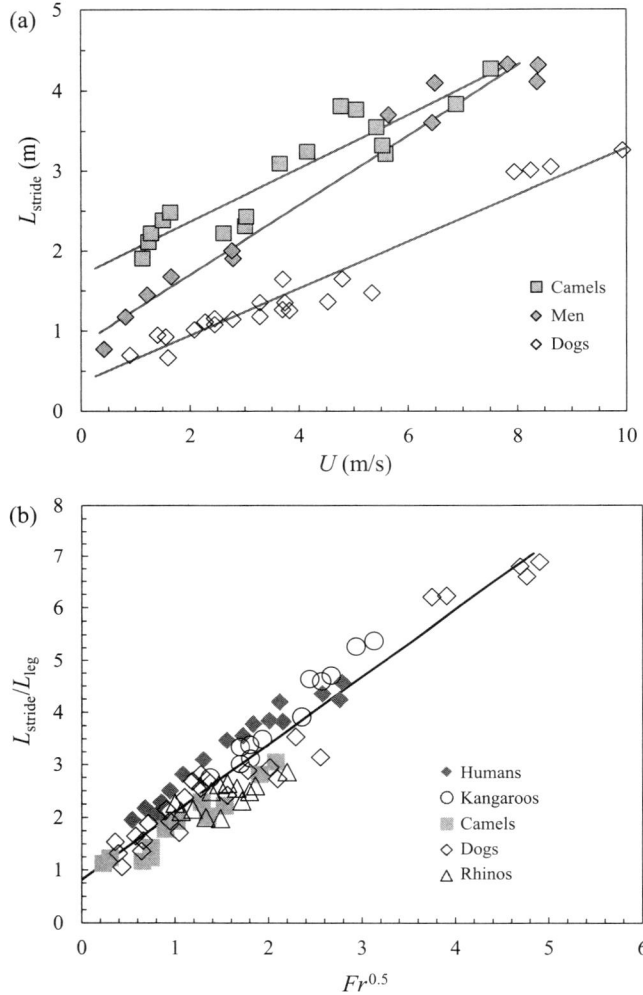

Figure 6.6
(a) Experimental data for two quadrupeds and human males for dimensional stride length versus running speed U. (b) Nondimensional stride length versus square root of Froude number. Here, data includes humans, kangaroos, camels, dogs, and rhinos. Each animal is represented several times (moving at different speeds). The dimensional data collapses fairly well onto a single straight line across species. Both figures were created by digitizing the data of plots shown by Alexander (1996). The lines in (a) are linear fits suggested by Alexander (1996).

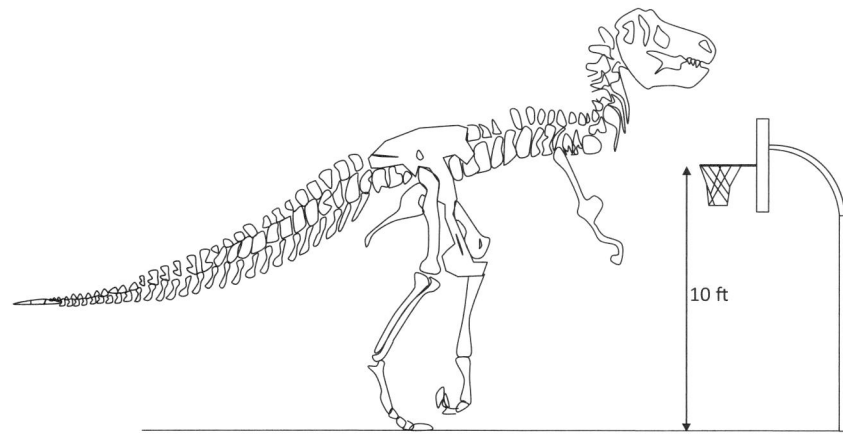

Figure 6.7
Sketch of a reconstructed skeleton of *Tyrannosaurus rex* shown next to a basketball hoop 3.05 m (10 ft) from the floor. A similar comparison and sketch is presented in the book by Alexander (1989).

Table 6.1
Estimated speeds of dinosaurs (adapted from Alexander, 1989)

	Estimated Leg Length (m)	Estimated speed (m/s)	Gait (mph)
data from Davenport Ranch (fig. 3.4 of Alexander)			
Large theropod	2.0	2.2	(4.9) walk
Small theropod	1.0	3.6	(8.1) run
Large sauropod	3.0	1.0	(2.2) walk
Large sauropod	1.5	1.1	(2.5) walk
data from Winton (fig. 3.1 of Alexander)			
Large theropod	2.6	2.0	(4.5) walk
Small theropods	0.13–0.22	3.0–3.5	(6.7–7.8) run
Ornithopods	0.14–1.6	4.3–4.8	(9.6–10.7) run

These types of estimates are presented in Alexander's book *The Dynamics of Dinosaurs* (Alexander, 1989). Table 6.1 is a reproduction of the associated estimates (see data from Alexander 1989, table 3.2).

Disappointingly (or encouragingly for prospective time travelers), the running speed of large dinosaurs was fairly low. For example, a large theropod such as *Tyrannosaurus rex* (see the sketch of a *T. rex* skeleton in figure 6.7) would have trouble catching a human casually jogging at 5 mph. Even small theropods like velociraptors are estimated to have run 8 mph. The fastest of the bunch were ornithopods, which seemed to max out near 11 mph.

Most healthy humans can muster running speeds of 15–20 mph. The human world speed record is roughly 25 mph. Presumably, your time-traveling self can likely outrun any dinosaur you meet. Just don't get nervous and trip.

Note that the estimate of hip height is not easy to make for all dinosaurs. There are only limited observations of detailed geometry, and Rainforth and Manzella (2007) point out some of the challenges with this.

6.6   Summary

- We began this chapter by proposing a set of rules ND1–ND6, which are useful in integrating physical intuition and experimental evidence with dimensional analysis. So far, we have discussed eleven rules in chapters 5 and 6. In chapter 11 we will present a twelfth rule and then summarize the twelve rules in a table.

- We explored particularly useful and historic examples of dimensional analysis where these rules helped facilitate a combination of dimensional analysis with physical intuition. The combination achieved surprisingly simple hypothetical functions for highly complex problems.

- We explored the spine patterns of liquid drops impacting a surface and derived a function for the nondimensional parameters. This derived scaling only partially collapses experimental data. We then discussed the hypothesis that this partial collapse is due to the effect of viscosity, which was neglected in the analysis. We then updated our hypothesis to include viscosity and saw that this necessarily increased the number of nondimensional parameters and the complexity of required experiments.

- We explored historical aspects of G. I. Taylor's analysis of the explosive yield of an atomic explosion. We showed how dimensional analysis leads us to the specific functional dependence of observed blast radius versus time but does not by itself lead to an estimate of the explosive energy yield. Nevertheless, experimental data does confirm the scaling law predicted by dimensional analysis and collapses this data.

- We developed an analysis for world-class weightlifters that helps compare world record lifts across weight classes. The analysis helped identify an index that potentially helps judge the quality of any one lifter and fairly well collapses the world record data, particularly for the bench press competition.

- We hypothesized variables involved in predicting the running speed of animals. The hypothesis was based on considerations of an inverted pendulum and its relevance to the transition between walking and running; and an additional parameter (stride length). Interestingly, the analysis ignored geometric similarity (e.g., was applied to bipeds and quadrupeds) but nevertheless yielded useful results.

Problems

6.1. Zohuri (2015) argues that the wave length λ of surface (liquid) waves can be described in terms of their frequency ω, wave height h, local average depth of the water d, liquid density ρ, gravitational acceleration g, and surface tension σ. The latter parameter being very important for very small waves.

    a. Propose a function for the nondimensional variables in this problem.

    b. For large waves, we might hypothesize that the wavelength will be insensitive to surface tension. Simplify your function to reflect this. To do this, apply rules ND1 and ND3 to first manipulate the parameters such that surface tension appears only in a single parameter. Then eliminate this parameter (as an example of rule ND6). A good choice for the latter, isolated parameter is the so-called Bond number $B_0 = \rho g \lambda^2 / \sigma$, which compares gravitation forces to surface tension forces.

6.2. Postulate a function for power generated by a wind turbine in a wind at velocity U. Clearly define your variables and identify their dimensions. Neglect the effect of wind viscosity and neglect compressibility effects. Also note that torque can be derived from power and angular velocity (cf. rule D1). Then use Ipsen's method for the dimensional analysis.

6.3. Repeat problem 6.2 but this time include the speed of sound as a parameter to characterize the importance of compressibility effects (i.e., high speeds). As the highest relevant velocity in the problem, choose the translational speed of the turbine blade tip.

7

Two Examples Combining Biomechanics and Fluid Mechanics

Walking on gravel is an uncomfortable experience for adults, but not for little children. A father who is twice as tall and 8 times as heavy as his 8-year-old daughter must support himself on feet whose surface area is only 4 times that of her feet. Thus, his "foot loading" is twice hers. No wonder he seems to be walking on hot coals.

—Hendrik Tennekes (2009)

This chapter presents further examples of dimensional analysis wherein we combine the rules proposed in chapters 5 and 6 to integrate physical insight. We consider two problems that involve animal powered propulsion: Olympic rowers and the dynamics of flight (both animal and machine fliers).

7.1 Olympic Rowers and a Word of Caution on the Experimental Validation of Scaling Laws

In the journal *Science*, Thomas A. McMahon (1971) asked the question: Why do larger rowboats with more rowers go faster than smaller boats with fewer rowers? How much faster should they go? He analyzed rowing shells to derive a law suggesting that rowing speed of geometrically similar shells will have a speed proportional to the number of rowers raised to the 1/9th power.

Figure 7.1 shows drawings similar to those of McMahon's paper, including the cross-sectional area of boat within the water and the beam (width) and length of boat.

McMahon's analysis was based on an analysis of approximate scaling relations. Hence, we will discuss his approach more clearly in chapter 13. Here we present a dimensional analysis that uses ideas leading to these approximations. The presentation will iterate between physical insights and dimensional analysis rules to try to capture the flavor of such approaches.

Further, I present a brief criticism of McMahon's experimental validation. Namely, I will argue that the functional dependence he predicts is extremely weak (a scaling to

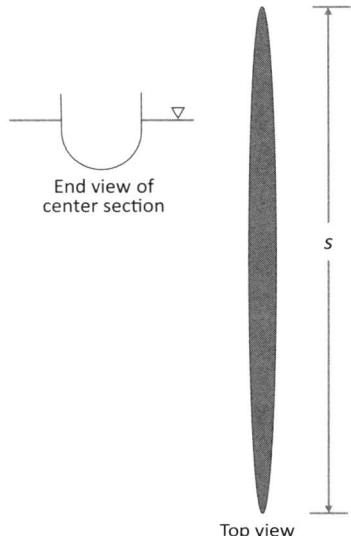

Figure 7.1
McMahon considered geometrically similar shells characterized by length s. He considered also the wetted area A of the shell under the water line; and n oarsmen each of weight W_0. McMahon (1971) presents a similar schematic describing his analysis.

the 1/9th power), and such a weak dependence is extremely difficult to validate with experiments. I then attempt to generalize this to the challenges of experimentally validating any power law dependence.

Physical insight 1: McMahon's analysis assumes each rower adds a constant weight W_0 but also contributes a constant power P_0 to rowing. McMahon also recognizes that the boat structures employed in these races get heavier for the number of rower seats, and so he assumes a constant ratio of boat structure weight W_{boat} to rower weight (so a two-rower boat is 2/3 the weight of a three-rower boat). We will also invoke McMahon's hypothesis that the drag on the boat will be some function of the wetted area of the boat (a concept we will explore further in section 10.3). We will characterize both the wetted area of the boat and its volume under the water in terms of a shell length at the water line s. Our approach is different from McMahon's in that we immediately invoke geometric similarity, so we have only one linear dimension. Also, McMahon formulates his arguments in terms of the entire boat length ℓ (as in his sketch). Despite this, his estimates are consistent with wetted/submerged geometry (e.g., McMahon's ℓ is a linear function of the number of rowers). Last, we anticipate that the buoyancy force of the submerged volume will be related to the weight of rowers (from Archimedes' principle). We postulate the following variables

Variable	Definition	Physical quantity
n	number of rowers	—
V	velocity of boat	LT^{-1}
ρ	density of water	ML^3
s	shell length (at the water line)	L
W_0	weight of each rower (a constant)	$M\,LT^{-2}$
P_0	power provided by each rower (constant)	$M\,L^2T^{-3}$
W_{boat}	weight of boat structure per rower (a constant)	$M\,LT^{-2}$
g	gravitational acceleration	LT^{-2}
μ	dynamic viscosity of water	$ML^{-1}T^{-1}$

Unlike McMahon, the current analysis is more general as we here include the dynamic viscosity of water in anticipation of the importance of drag on this problem. Again, our geometric similarity concerns the geometry under the water line, and we assume both wetted area and submerged geometries are characterized by the single length scale s. We hypothesize a function as follows in terms of a single length scale s as per rule D5:

$$n = f_1(V,\ \rho, s, P_0, W_0, W_{boat}, g, \mu).$$

n is nondimensional but not constant; hence, we cannot apply rule D3. We apply Ipsen's method as follows:

Eliminate L with s:

$$n = f_2\left(\frac{V}{s},\ \rho s^3,\ \frac{P_0}{s^2},\ \frac{W_0}{s},\ \frac{W_{boat}}{s},\ \frac{g}{s},\ \mu s\right).$$

— T^{-1} M MT^{-3} MT^{-2} MT^{-2} T^{-2} MT^{-1}

Eliminate M with ρs^3:

$$n = f_3\left(\frac{V}{s},\ 1,\ \frac{P_0}{\rho s^5},\ \frac{W_0}{\rho s^4},\ \frac{W_{boat}}{\rho s^4},\ \frac{g}{s},\ \frac{\mu}{\rho s^2}\right).$$

— T^{-1} — T^{-3} T^{-2} T^{-2} T^{-2} T^{-1}

Eliminate T with s/V:

$$n = f_5\left(1,\ 1,\ \frac{P_0}{\rho s^2 V^3},\ \frac{W_0}{\rho s^2 V^2},\ \frac{W_{boat}}{\rho s^2 V^2},\ \frac{sg}{V^2},\ \frac{\mu}{\rho V s}\right)$$

—   —   —   —   —   —   —   —

or

$$n = F_1\left(\frac{P_0}{\rho s^2 V^3},\ \frac{W_0}{\rho s^2 V^2},\ \frac{W_{boat}}{\rho s^2 V^2},\ \frac{sg}{V^2},\ \frac{\mu}{\rho V s}\right).$$

Dimensional analysis has reduced the problem from nine to six variables.

Let us now apply further physical intuition. To apply rule ND5, we assume the weight ratio W_{boat}/W_0 is constant. Accordingly, we can force the nondimensionalization to include this ratio by applying rule ND1 to divide the third nondimensional parameter in the parenthesis by the second without loss of generality, hence

$$n = F_2\left(\frac{P_0}{\rho s^2 V^3}, \frac{W_0}{\rho s^2 V^2}, \frac{W_{boat}}{W_0}, \frac{sg}{V^2}, \frac{\mu}{\rho V s}\right).$$

Note this does not change the number of parameters in the formulation. Since W_{boat}/W_0 is a constant for this problem, we can simply absorb it as per rule ND5 into the definition of the function (since it is a nondimensional constant). Hence,

$$n = F_3\left(\frac{P_0}{\rho s^2 V^3}, \frac{W_0}{\rho s^2 V^2}, \frac{sg}{V^2}, \frac{\mu}{\rho V s}\right).$$

Next, we recognize the relation of the last two terms to Froude number and Reynolds number, respectively. Using rule ND1, we raise the penultimate parameter to the negative ½ power and invert the last to show this explicitly.

$$n = F_3\left(\frac{P_0}{\rho s^2 V^3}, \frac{W_0}{\rho s^2 V^2}, \frac{V}{\sqrt{sg}}, \frac{\rho V s}{\mu}\right).$$

As we shall discuss in chapter 9, the drag force on a boat is related to both Froude number and Reynolds number. We can express this in terms of nondimensional parameters as

$$C_D = G\left(\frac{V}{\sqrt{sg}}, \frac{\rho V s}{\mu}\right).$$

At this point, we invoke rules ND2 and ND4 to recognize that the dependence on Reynolds and Froude numbers will be only through the drag coefficient, and so we replace these as

$$n = F_4\left(\frac{P_0}{\rho s^2 V^3}, \frac{W_0}{\rho s^2 V^2}, C_D\right).$$

Physical insight 2: The shells used in competitions all have similar shapes and travel at very similar velocities (varying by, say, 10%–20%). Hence, we will assume all of these boats have approximately the same drag coefficient. This enables us to invoke rule ND5 and rewrite an approximation for our function such that we absorb the nondimensional value of the roughly constant drag coefficient into the function as:

$$n = F_5\left(\frac{P_0}{\rho s^2 V^3}, \frac{W_0}{\rho s^2 V^2}\right).$$

We should recognize that the first term in the function approximately has the form of input power per power dissipated by drag (for an assumption of constant drag coefficient). That is, for constant drag coefficient, we can write the denominator as $(\rho s^2 V^2 C_D)V$, which is drag force multiplied by velocity. Hence, we use rule ND1 to multiply the first term by nondimensional n to include total power as follows:

$$n = F_6\left(\frac{nP_0}{\rho s^2 V^3}, \frac{W_0}{\rho s^2 V^2}\right).$$

Physical insight 3:   Taking a cue from Archimedes' principle, we can force the formulation to include a ratio of total weight of the boat and rower(s) system and the weight of the (geometrically similar) displaced water volume. To do this, we can group the total weight of the boat and rowers' weight $n(W_0 + W_{\text{boat}})$ with a weight of displaced volume directly proportional to $\rho s^3 g$ (basically the principle of rule D1). This is a choice resulting from knowledge of the physics and not dimensional analysis alone. To force this, we can multiply the second term in the function by the square of a nondimensional in this problem, the square of Froude number, $V^2/(sg)$. In this second term, we accordingly also replace the partial weight, W_0, with the total weight $n(W_0 + W_{\text{boat}})$, as per rule D1. This results in the following:

$$n = F_7\left(\frac{nP_0}{\rho s^2 V^3}, \frac{n(W_0 + W_{\text{boat}})}{\rho s^3 g}\right).$$

Physical insight 4:   We make the assumption that the number of rowers only affects the problem in the fact that rowers contribute to total power (via nP_0) and by increasing the weight and displaced volume (via their collective contribution to the increased total displaced water volume $\rho s^3 g$). Accordingly, we drop n as an independent parameter and keep it only in the other relevant terms (invoking rule ND2 for a known dependence):

$$1 = F_8\left(\frac{nP_0}{\rho s^2 V^3}, \frac{n(W_0 + W_{\text{boat}})}{\rho s^3 g}\right).$$

Physical insight 5:   Last, Archimedes' principle applies strictly to the hydrostatic case of nonmoving boats and no fluid flow. However, we will assume that the boats move slowly enough that Archimedes' principle approximately applies (i.e., their motion is unlike skiing or surfing). This last hypothesis helps us approximate the argument of the function as a constant. Since the function of a constant is constant (rule ND5, cf. section 2.4):

$$\frac{nP_0}{\rho s^2 V^3} = F_9\left(\frac{n(W_0 + W_{\text{boat}})}{\rho s^3 g}\right) = \frac{nP_0}{\rho s^2 V^3} = F_9\left(\frac{n(W_0 + W_{\text{boat}})}{\rho s^3 g}\right) \cong F_9(a_0) = \text{constant} = a_1,$$

where

$$a_0 \equiv \rho g s^3 /(n(W_0 + W_{\text{boat}})) \Rightarrow n = \frac{\rho g s^3}{a_0(W_0 + W_{\text{boat}})}. \tag{7.1}$$

For the second constant:

$$\frac{\rho V^3 s^2}{n P_0} = a_1 \Rightarrow s^2 = \frac{a_1 n P_0}{\rho V^3} \Rightarrow s = \left(\frac{a_1 n P_0}{\rho V^3} \right)^{1/2}. \tag{7.2}$$

We now combine equations (7.1) and (7.2) to eliminate the length coordinate and obtain

$$n = \frac{g}{a_0(W_0 + W_{\text{boat}})} \left(\frac{a_1 n P_0}{V^3} \right)^{3/2} \Rightarrow V^{9/2} = \frac{g(a_1 P_0)^{3/2} n^{1/2}}{a_0(W_0 + W_{\text{boat}})}$$

$$\Rightarrow V = \left(\frac{g(a_1 P_0)^{3/2}}{a_0(W_0 + W_{\text{boat}})} \right)^{2/9} n^{1/9}$$

or

$$V = a_2 n^{1/9},$$

where

a_2 = constant.

The dimensional analysis predicts that the velocity of the boat should approximately scale as the number of rowers raised to the 1/9th power! This is a very weak dependence. The hypotheses and reasoning seem sound, but can we show that this is accurate? We explore this using figure 7.2, which shows data from the world record times, T, to finish 2000 m races as a function of rowers. We explore the scaling by plotting the data as $T \propto n^{-1/m}$ for several values of the integer m. We see McMahon's theory ($m = 9$) has reasonable agreement with these world records. However, we note that scaling laws of $m = 8$ or $m = 10$ are just as reasonable given the limited range of the data.

The ambiguity among the various scalings in figure 7.2 helps identify a couple of potential challenges of dimensional analysis and power law data collapses. The analysis showed an elegant scaling law of the form $T = $ constant $n^{-1/9}$, since this scaling collapses the data. However, true validation would require a sufficient range of experimental data to discriminate among the various theories. In this case, we simply do not have enough variation in the experimental data. This is especially true here as the power dependence is extremely weak. Consider we have an 8:1 variation (just under one order of magnitude) in n. The 9th root of 8 is just 0.9, or just a 10% drop in T. I estimate we would need experiments with boats rowed by up to about 80 world-class rowers (!) before we can validate this power law. In McMahon's defense, there are simply no 80-rower boat events in the

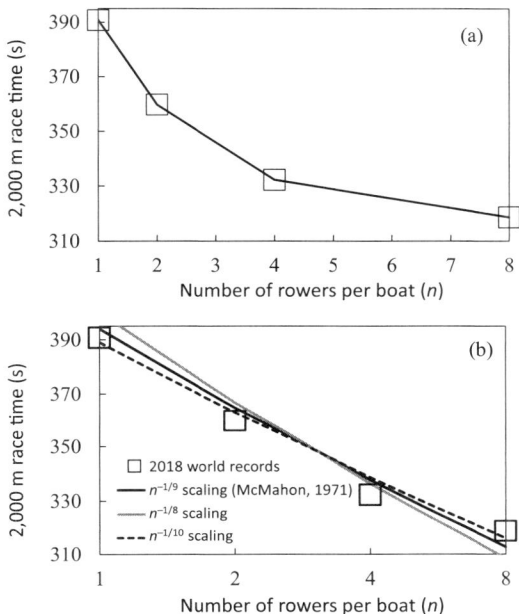

Figure 7.2
World record times for 2000 m rowing races for boats with $n = 1, 2, 4,$ and 8 rowers. The top plot (a) is a plot showing times T versus rower number. The bottom plot (b) shows a comparison between times T and the logarithm of the number of rowers. Shown together with the data in (b) are power laws of the form $T = $ constant $n^{-1/m}$ with m equal to 8, 9 (McMahon), or 10. Given the limited variation of boat rowers (8:1), it is difficult to experimentally validate the precise scaling law.
Source: Data obtained from http://www.concept2.com (accessed July 2018).

Olympics. Still, the analysis is an excellent example of reduction of variables using physical intuition.

Last, note that another (and perhaps more intuitive) way of tackling this boat and rower problem is to directly write approximate algebraic equations for Archimedes' principle, a drag law, and so forth and then perform dimensional analysis on these approximate algebraic equations. We shall demonstrate this approach in detail in chapter 13 when we look at scaling based on approximate governing equations. We shall also use this approach to more explicitly demonstrate the dependences of boat velocity on drag coefficient, rower power, rower weight, and boat weight.

7.2   A Derivation for the Great Flight Diagram: Approximate Data Collapse

Consider all of the complex dynamics associated with the flights of animals and machines of all shapes and sizes. Fliers in Earth's atmosphere include gnats and houseflies, small and large birds, ultralight airplanes, F-16 fighter planes, and Boeing 747s. Tennekes (1997) presents an intuitive and inspiring description of the concept of flight. He uses basic concepts of density and velocity to describe mass flow deflected by wings and subsequently the deflection and change of momentum flow created by wings. He attempts to unify the dynamics of all flying objects, from a midge (a small insect) to a jumbo jet.

We here present a dimensional analysis for the flight problem that will follow the format we have used so far. Importantly, unlike Tennekes, we will include in our analysis the dynamic viscosity of air. As we shall see, this generalization leads us to derive a clear Reynolds number dependence for the problem. This in turn will let us more clearly introduce drag and lift coefficients. Hence, the current approach is different from Tennekes's, but I believe the approach is consistent with his ideas.

Our goal is to relate flight velocity to flyer weight mg, and so we choose the following variables:

Variable	Definition	Physical quantity
U	velocity of level flight	LT^{-1}
s	wingspan of flier	L
M	mass of flier	M
g	gravitational acceleration	LT^{-2}
μ	dynamic viscosity of air	$ML^{-1}T^{-1}$
ρ	density of air	ML^{-3}

Again, we include the dynamic viscosity of air in anticipation of the importance of drag and lift for this problem. Interestingly, in choosing a single length scale s to describe all fliers, we will implicitly assume a basic geometric similarity for all fliers. Although Tennekes does not discuss this geometric similarly explicitly, we might here consider this basic shape a single torso (a fuselage) with two major symmetric wings on either side. The wings tend to have more area than the central shape and extend in a direction approximately perpendicular to the free stream direction. This bold approximation is depicted schematically in figure 7.3.

From this concept of geometric similarity and the postulated variables, we can hypothesize a desired function of the form:

$$U = f_1(m, g, s, \rho, \mu). \tag{7.3}$$

Here we have (blatantly and roughly) assumed geometric similarity across all flyers and, as per rule D5, use a single length scale to describe them all (witness the audacity of Figure 7.3). We explicitly include power of flight as per rule ND4, and we will discuss this further in section 7.2.1. We apply Ipsen's method. Eliminate L with s:

$$\frac{U}{s} = f_2\left(m, \ \frac{g}{s}, \ 1, \ \rho s^3, \ \mu\right).$$

$$T^{-1} \qquad M \ \ T^{-2} \ - \ M \ \ MT^{-1}$$

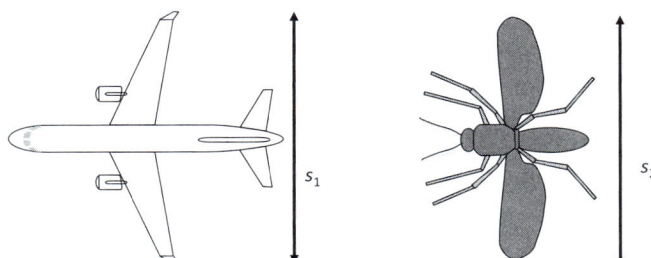

Figure 7.3
Two very roughly geometrically similar flyers characterized by a single dimension (say, wingspan s). The shapes can perhaps be described as an elongated cylindrical body with two major symmetric wings extending on either side. Neither commercial jet nor midge is shown to scale!

Eliminate M with ρs^3:

$$\frac{U}{s} = f_3\left(\frac{m}{\rho s^3}, \frac{g}{s}, \quad 1, \quad 1, \quad \frac{\mu}{\rho s^3} \right).$$

$$T^{-1} \qquad - \quad T^{-2} \; - \; - \quad T^{-1}$$

Eliminate T with s/U:

$$1 = F_1\left(\frac{m}{\rho s^3}, \frac{gs}{U^2}, \frac{\mu}{\rho Us} \right). \tag{7.4}$$

$$- \qquad - \qquad - \qquad -$$

So we arrive at three nondimensional parameters, one of which is an inverse Reynolds number.

Physical insight 1: As discussed in chapters 4 and 9, we expect (nondimensional) drag coefficient, C_D, and lift coefficient, C_L, each to be a function of Reynolds.

$$C_D = f(Re)$$

$$C_L = f(Re).$$

Here, we will make a very bold (i.e., risky) assumption to derive Tennekes's result: Namely, we will assume that C_D and C_L are both constants. If so, we can invoke rule ND5 and absorb these assumed constant parameters into a newly defined function as

$$\frac{gs}{U^2} = F_2\left(\frac{m}{\rho s^3} \right).$$

Physical insight 2: The dependent variable in the latter function is just the mass of the flier divided by the mass of the displaced air. For geometrically similar fliers in Earth's (air) atmosphere, we can assume this nondimensional parameter is approximately constant, hence

$$\frac{gs}{U^2} = F_3\left(\frac{m}{\rho s^3}\right) = \text{constant} = c_0. \tag{7.5}$$

Hence, we can write equation (7.5) as

$$g = c_0 \frac{U^2}{s}.$$

To force this equation to include the weight of the flyer and parameters from the problem, we can multiply the left-hand side by the constant value $m/\rho s^3$ and absorb this value into a new constant:

$$\frac{mg}{\rho s^3} = c_1 \frac{U^2}{s}$$

or

$$mg = c_1 \rho U^2 s^2. \tag{7.6}$$

This relation can be interpreted as the condition for level flight where weight is equal to lift. Lift on the right-hand side in turn can be interpreted as a constant lift coefficient (here c_1) and the product of dynamic pressure and a planform area of the flyer. That is,

$$\frac{mg}{\rho U^2 s^2} = \frac{\text{gravitational force}}{\text{lift force (due to dynamic pressure)}}. \tag{7.7}$$

We wish to obtain a scaling between weight and velocity. To eliminate length scale, we should relate s to the flyer mass. Since we consider geometrically similar flyers, we can express the mass as a function of the flyer density, ρ_f, a volume-averaged density of the flyer's body/structure. For bodies of jumbo jets to houseflies, this volume-averaged value should not deviate much from the density of water (e.g., consider YouTube videos of airliners slowly and purposely sunk into ocean water). Hence, we can express mass as follows:

$$m = c_2 \rho_f s^3 \text{ or } mg = c_2 \rho_f s^3 g. \tag{7.8}$$

Dividing the two expressions (equations [7.6] and [7.8]) for weight:

$$1 = \frac{c_1 \rho U^2 s^2}{c_2 \rho_f s^3 g} \text{ or } s = \frac{c_1 \rho U^2}{c_2 \rho_f g}.$$

We can use this new expression for length scale s to eliminate s from the level flight condition:

$$mg = c_1 \rho U^2 s^2 = c_1 \rho U^2 \left(\frac{c_1 \rho U^2}{c_2 \rho_f \, g} \right)^2 = c_1 \rho \left(\frac{c_1 \rho}{c_2 \rho_f \, g} \right)^2 U^6 = c_3 U^6. \tag{7.9}$$

The brazen assumption of geometric similarity, the ruthless pairing down of variables to a small list, and the approximations of approximately constant quantities (including lift coefficient!) provides a beautifully simple scaling law between flyer weight and velocity. The next section compares this scaling to experimental data.

7.2.1 The Great Flight Diagram

The evidence that the scaling of equation (7.9) holds some insight is in its comparison with experimental data. Tennekes (1997) presents the plot shown in figure 7.4, which he has called the great flight diagram.

From our assumptions, we see that likely the most restrictive assumption (and an important reason for the scatter in the diagram) is the fact that we have neglected dependence of lift and drag coefficients on Reynolds number (and so kinematic viscosity). We will discuss this further in the next section.

Tennekes discusses this figure in detail, and we list here a few colorful examples from his discussions.

- The animals to the left of the vertical line (at 10 m/s, or about 22 mph) might not be able to fly to their nest/home in a strong wind.

- The data point marked Pteranodon is an estimate for an extinct flying reptile, the largest of the Cretaceous era (weighing 170 N with a 7 m wingspan). It's heavier per flight velocity than the trend line, but it is believed to have spent most of its flight time soaring above cliffs (e.g., not continuously flapping).

- The data point marked *Argentavis magnificens* is an estimate for the largest flying animal in history, which is also believed to have mostly soared. This giant bird weighed 700 N and had a 7 m wingspan and wing area of 8 m^2. The estimates for larger soaring birds deviate from the line to a larger degree than smaller birds (which tend to flap their wings as they fly).

- Recall that $W = c_3 U^6$. Human power per weight is relatively low, and so the Gossamer Condor human-propelled plane has very large wings and very low velocity. This very light vehicle is powered by a single pedaling pilot struggling to stay aloft. The situation also affects the performance of the solar-powered aircraft shown in the data.

- The power versus weight versus velocity issue affects all fliers. For example, sports planes (leisure type craft) tend to be slower (above and to the left of the trend line), but they creep toward the trend line as engine power increases.

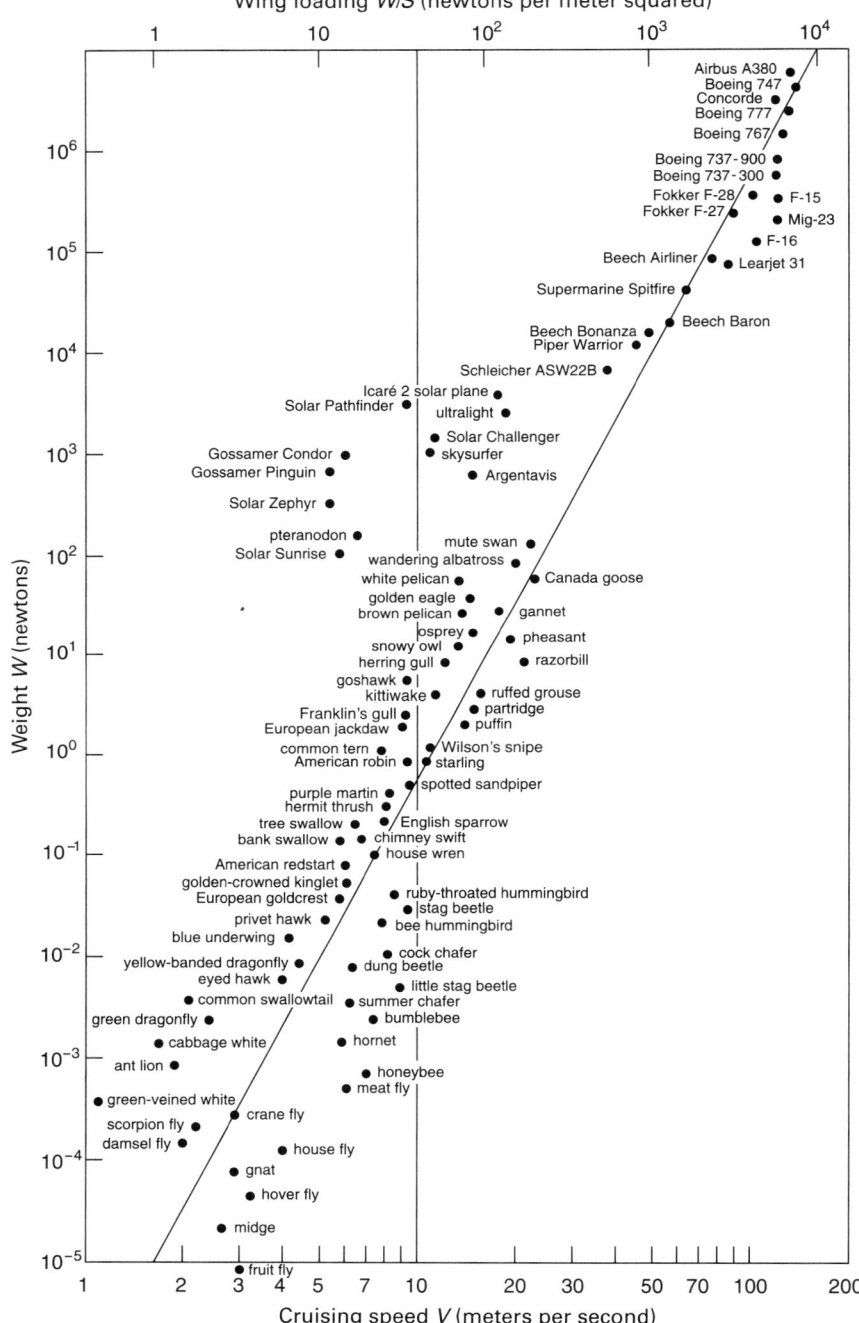

Figure 7.4
Tennekes's great flight diagram is a plot of weight W of a wide range of fliers versus cruising speed for horizontal flight U. The line is a power law of the form $W = $ constant U^6, where the value of the constant is set at 47 for these axes.

- Historically, aircraft engines gained greatly in power per weight ratio. Low power initial implied low cruising speeds (one advantage being short takeoff and landing, e.g., on grass fields). As engine power increased through World War I and World War II, total weight also increased. This greater power enables higher velocity (to defy drag) and so requires less wing area. Interestingly, the net result is that wing area is roughly constant for increases in engine power.

- For the smallest fliers, the simple aerodynamic principles applied in the derivation do not apply since their characteristic Reynolds number is so low. The poor fruit fly's data point is off the plot. Tennekes points out the fruit fly's flight is in a low Reynolds number (viscous) regime, and he imagines its flight is like "flying in syrup."

- The very top speed of the fliers is capped at about 250 m/s by the local speed of sound (about 295 m/s at 30,000 ft). Transonic and supersonic flight requires analysis of compressibility effects (e.g., using Mach number), which were not considered here.

The approximate collapse of data for fliers across 11 orders of magnitude of weight is nevertheless remarkable.

7.3 Justification for Excluding Flier Power from the Formulation

Recall our ruthless pairing down of the list of salient variables in this problem and its associated hypothetical function for flier velocity:

$$U = f_1(m, g, s, \rho, \mu).$$

At first look, it would seem useful (even prudent) to include an additional variable: Power. Flier power, P, helps determine the ability of a flier to overcome drag force, F_D, which is itself a strong function of velocity (e.g., drag scales as U^2 for approximately constant drag coefficient). Also, the absolute output power of a flyer (e.g., in derived units of watts) should somehow be limited by its mass. We here discuss a brief justification for excluding power of the flier, which helps emphasize our very important rule D2: namely, to exclude variables that can be uniquely derived from (or expressed as functions of) other variables in the problem. In consideration of whether to include power of the flyer, we formulate the power dissipated by the flight motion as follows:

$$P = F_D U.$$

We then express exactly in terms of a drag coefficient as a function of Reynolds number (see section 4.4) as follows:

$$P = \frac{1}{2} \rho U^3 C_D (Re), \text{ where } Re = \frac{\rho U s}{\mu}.$$

Hence, we see that power is derivable from variables already included in the problem (namely ρ and U and Reynolds number) and so we can exclude it. That is, we invoked rule D2 and exclude variables for lift and power as they will be approximately determined from the included quantities.

7.4   Including Viscosity in the Flight Analysis

In this section, I present a generalization of the Tennekes's flight analysis, which explicitly includes the effect of viscosity (and so Reynolds number). We first use dimensional analysis alone to explicitly derive and show the Reynolds number dependence. We then use the concept of a lift coefficient (itself a result of dimensional analysis!) to derive a simple expression for the Reynolds number dependence.

Recall that dimensional analysis alone led to the following function (equation [7.4]) relating the hypothesized relevant variables in the problem

$$1 = F_1 \left(\frac{m}{\rho s^3}, \ \frac{gs}{U^2}, \ 1, \ 1, \ \frac{\mu}{\rho U s} \right).$$

Applying rule ND1, we can rewrite this as

$$\frac{gs}{U^2} = G_1 \left(\frac{m}{\rho s^3}, \ \frac{\rho U s}{\mu} \right).$$

And so we have forced the problem to explicitly include a traditional Reynolds number of the form $\rho U s / \mu$. Given this form, we can still apply physical insight 2 of section 7.2 and assume that all flyers have approximately the same mass density. In doing so, we use rule ND5 to absorb the approximately constant nondimensional term $m/\rho s^3$ into the definition of a new function as

$$\frac{gs}{U^2} = G_2 \left(\frac{\rho U s}{\mu} \right) = G_2(Re). \tag{7.10}$$

The expression now explicitly shows that the relation between the length scale and speed of flyers explicitly depends on Reynolds number.

To show the Reynolds number dependence explicitly in the final result, we invoke two of the aforementioned physical insights. First, we again make the assumption that Earth's flyers have approximately the same density leading the earlier defined constant c_2 (as per equation [7.8]):

$$c_2 = \frac{mg}{\rho_f s^3 g}. \tag{7.11}$$

Second, we recognize that the weight of the flyer scales as a lift force that is related to the dynamic pressure. However, this time, consistent with keeping the effect of viscosity in the analysis, we do not assume that the lift force is directly proportional to the dynamic pressure, but instead express the lift force in terms of a lift coefficient (which depends on Re, discussed earlier). Accordingly, we modify equation (7.7) as follows:

$$\frac{mg}{\rho U^2 s^2 C_L} = \frac{\text{gravitational force}}{\text{lift force}} = 1. \tag{7.12}$$

Here, the ratio is exactly unity (in level flight) by definition of the lift coefficient. In other words, the variable lift coefficient C_L has here replaced our earlier constant c_1, making the expression much more general. Including a lift coefficient implies that the lift force does not necessarily scale with dynamic pressure alone, as we had assumed earlier. We can combine equations (7.11) and (7.12) as follows:

$$c_2 = \frac{mg}{\rho_f s^3 g} \underbrace{\frac{\rho U^2 s^2 C_L}{mg}}_{1} = \frac{\rho U^2 C_L}{\rho_f gs}. \tag{7.13}$$

Rearranging equation (7.13),

$$s = \frac{\rho U^2 C_L}{\rho_f g c_2}. \tag{7.14}$$

Last, we substitute equation (7.14) back into equation (7.12) to derive simply

$$\frac{mg}{\rho U^2 C_L} \frac{1}{s^2} = \frac{mg}{\rho U^2 C_L} \frac{(\rho_f g c_2)^2}{(\rho U^2 C_L)^2} = mg \frac{(\rho_f g c_2)^2}{(\rho C_L)^3} \frac{1}{U^6} = 1$$

or

$$mg = \frac{(\rho C_L)^3}{(\rho_f g c_2)^2} U^6.$$

Last, to explicitly show the dependence of the lift coefficient on viscosity (and therefore Reynolds number Re), we write this as

$$mg = \frac{(\rho C_L(Re))^3}{(\rho_f g c_2)^2} U^6.$$

Here the notation $C_L(Re)$ refers to the variable lift coefficient's dependence on Reynolds number. Again, the analysis captures the scaling with the 6th power of velocity. However, this time our result is more general than that of Tennekes, inasmuch as it includes explicitly the effect of Reynolds number on the scaling. The value of Re varies widely in the data of the great flight diagram, and this helps, at least in part, explain the significant variations from a single-valued function observed in the data. We can think of the scatter in the great flight diagram as evidence strongly suggesting that the analysis is missing an important variable, namely viscosity.

7.5 Summary

- We used dimensional analysis to relate the number of rowers to the speed of a row boat. We assumed geometrically similar shapes submerged under the waterline and assume all drag came from water forces on the boat. Dimensional analysis alone resulted in a six-dimensional problem. We applied physical insights and approximations to reduce this to a simple scaling law.

- An important fluid mechanics concept for the rowing problem was the assumption that boat drag coefficient should be a function of Reynolds and Froude number. For the relatively narrow range of boat speeds for competitive rowers, we assumed that the drag coefficient was a constant value.

- McMahon's scaling predicts that boat times will scale with the number of rowers as $n^{-1/9}$ dependence. This very weak dependence is difficult to validate given the experimental data, and it highlights the importance of having sufficient variation in experimental parameters to demonstrate power laws.

- An important biomechanical concept for the rowing problem was that each rower influences the problem only by contributing an additive contribution to the total input power and additive contribution of weight (which increases drag via the geometrically scaled geometry below the waterline).

- We looked at formulating the flight speed U of fliers as a function of their weight mg. Our presentation included a dimensional analysis that combined physical intuition with ruthless and bold approximations including constant flier density and geometric similarity type assumption (for airliners and mosquitos!). We also assumed constant values of both the lift and drag coefficients.

- For earthly fliers, we derived a remarkably simple but still useful relation of the form: $mg \propto U^6$. Amazingly, $mg \propto U^6$ shows some significant collapse of data across 11 orders of magnitude of weight and two orders of magnitude of velocity.

- Last, we derived a more generalized version of Tennekes's scaling law, which explicitly includes the effect of viscosity via a variable lift coefficient and its explicit dependence on Reynolds number. We hypothesized that the latter formulation helps explain some of the scatter of the data about the predicted scaling law.

Problems

7.1. Consider again the scaling derived here between the number of rowers and the speed (and race times) of boats.

a. For this scaling, estimate the number of required rowers required in a geometrically similar hull to row at 2 and 3 times the velocity of a single rower.

b. Estimate the same number of rowers but for a scaling of the form $T = $ constant $n^{-1/8}$. These estimates should give an intuition for the wide variation in rower number required to discriminate between the two scaling laws, considering experimental uncertainty, and so on.

7.2. Given the scaling we derived for rowing, estimate the competitive edges gained by either one of the following: a 5% reduction in drag coefficient or a 5% reduction in total weight. Compare your estimated time reductions to the time differences among the first, second, and third place 2016 Olympic Games, Men's single sculls. The times were as follows:

Medal	Rower	Time
Gold	Mahe Drysdale (New Zealand)	6:41.34
Silver	Damir Martin (Croatia)	6:41.34
Bronze	Ondrej Synek (Czech Republic)	6:44.10

Source: http://nbcolympics.com

7.3. Note the secondary abscissa along the top of the great flight diagram. This axis relates wing loading in units of pressure (Newtons per square meter) to flight speed across the data. Derive an expression for such wing loading consistent with Tennekes's scaling analysis.

7.4. Set up and perform a dimensional analysis applicable to the quotation from Tennekes at the beginning of this chapter. Relate foot surface pressure to the height of the walker.

8

Buckingham Pi Theorem: An Alternate Method

> There must be as many different units as there are different kinds of quantities to be measured, but in all dynamical sciences it is possible to define these units in terms of the three fundamental units of Length, Time, and Mass.
>
> —James Clark Maxwell and Joseph John Thompson

The Buckingham Pi theorem is of historic significance and the most commonly taught method of performing dimensional analysis. For example, it is the primary method taught in most undergraduate courses in fluid mechanics (e.g., White, 2016; Fox et al., 2015). Such courses are often the only place where undergraduate engineering students learn about dimensional analysis. A minor advantage of the Buckingham Pi theorem is that some of the process lends itself to teaching students about matrix algebra. We here present it only for completeness and as a comparison to Ipsen's method. We end this chapter with some notes comparing the two methods.

8.1 Buckingham Pi Theorem

The following describes the Buckingham Pi theorem process. Given a relations among n-dimensional variables of the form

$$q_1 = f(q_2, q_3, \ldots, q_n)$$

or alternately

$$g(q_1, q_2, q_3, \ldots, q_n) = 0,$$

then n parameters may be grouped into $(n-m)$ independent ratios of the variables q (called Π groups), which are related as

$$\Pi_1 = F(\Pi_2, \Pi_3, \ldots, \Pi_{n-m})$$

or

$$G(\Pi_1, \Pi_2, \Pi_3, \ldots, \Pi_{n-m}) = 0.$$

Here, m is the rank of the so-called dimensional matrix describing the dimensions of the variables q. This value will be described further later.

> Comment: Usually, m is the minimum number of independent dimensions required to specify the dimensions of all variables $q_1, q_2, \ldots q_n$. In a few cases (e.g., when different dimensional systems are used to express the variables), this is not the case.

Process for finding Π groups:

1. List all of the parameters involved in the problem.

2. Select a set of fundamental/primary dimensions (e.g., M, L, T).

3. List the dimensions of all parameters in terms of fundamental/primary dimensions (e.g., in terms of M, L, and T). Construct the dimensional matrix. Columns are relevant parameters (e.g., force, velocity, pressure, etc.). Rows are the fundamental dimensions. The rank, m, of the dimensional matrix is the dimension of the largest nonzero (square) determinant (e.g., if there is a nonzero three-by-three submatrix, then the rank is three or larger).

4. Select from the list of parameters a number of repeating parameters equal to m. The repeating parameters should contain all of the primary dimensions and not have primary dimensions that are simply powers of another repeating parameter (e.g., don't choose a length and a volume). Importantly, the repeating parameters should not form a nondimensional variable among themselves.

5. Set up the dimensional equations, combining the repeating parameters selected in step 4 with each of the remaining parameters, to form dimensionless groups.

6. Check to see that each group is in fact dimensionless.

8.2   Pressure Drop in a Pipe Explored Using the Buckingham Pi Theorem

We here present an example of Buckingham Pi theorem. Consider the pressure drop inside a long, horizontal, cylindrical tube (a pipe). Fluid mechanicians have long known (from experimental observations) that this pressure drop is a function of the characteristic length scale of the roughness of the inside wall of the tube (White, 2016). The formulation of the problem presented here is adapted from a similar example provided by Fox et al. (2015) with modifications in the format of the presentation.

1. Postulate the key variables as follows:

Variable	Definition	Physical quantity
Δp	pressure drop in pipe (dependent variable)	$ML^{-1}T^{-2}$
U	area-averaged velocity through pipe	LT^{-1}
d	inner diameter of pipe	L
s	length of pipe	L

ε	roughness element length (e.g., standard deviation)	L
ρ	density of fluid	ML^3
μ	dynamic viscosity of fluid	$ML^{-1}T^{-1}$

2. Select a set of fundamental/primary dimensions (e.g., M, L, T).

3. List variables in terms of primary dimensions.

$$\Delta p = \qquad f_1(U, \ d, \ s, \ \varepsilon, \ \rho, \qquad \mu).$$
$$ML^{-1}T^{-2} \quad LT^{-1} \ \ L \ \ L \ \ L \ \ ML^{-3} \ ML^{-1}T^{-1}$$

Next, construct the so-called dimensional matrix. The rows of the matrix correspond to the fundamental dimensions (here M, L, and T). The columns are the variables of interest. The elements of the matrix are the power to which the fundamental dimension is raised within the variable's dimension as follows. Note that the order of rows and columns is arbitrary. Hence, we list below two example versions of the dimensional matrix.

Version 1 of dimensional matrix:

	Δp	μ	ρ	U	s	ε	d
M	1	1	1	0	0	0	0
L	−1	−1	−3	1	1	1	1
T	−2	−1	0	−1	0	0	0

Version 2 of dimensional matrix:

	Δp	U	s	ε	ρ	μ	d
M	1	0	0	0	1	1	0
L	−1	1	1	1	−3	−1	1
T	2	−1	0	0	0	−1	0

The rank, m, of the dimensional matrix is the dimension of the largest nonzero (square) determinant (e.g., if there is a nonzero three by three submatrix, then the rank is three or larger). These square matrices can be constructed by putting rows or columns in any order (and technically, all such combinations must be explored!). We'll here first check the leftmost square matrix within version 1. Note that, in anticipation of this, we wrote version 1 such that most of the powers expressed in it are nonzero (although this, of course, does not ensure the determinant will be nonzero). Continuing, the leftmost square matrix of version 1 matrix is:

$$\begin{matrix} 1 & 1 & 1 \\ -1 & -1 & -3 \\ -2 & -1 & 0 \end{matrix}$$

For which the determinant D is

$$D = \begin{vmatrix} 1 & 1 & 1 \\ -1 & -1 & -3 \\ -2 & -1 & 0 \end{vmatrix} = 1(-1-3)-1(-1-6)+1(1-2) = 2.$$

The determinant is nonzero and so, the rank of our dimensional matrix is 3 (equal to rank of highest nonzero square submatrix). There are no larger (e.g., four by four) square matrices within the dimensional matrix, so we can stop looking.

4. Select from the list of parameters a number of repeating parameters equal to m. We here choose

   $\rho, U, d.$

   These repeating parameters contain all primary dimensions and do not have primary dimensions that are powers of another repeating parameter. Importantly, these repeating parameters should also not form a nondimensional group among themselves. Note that the latter must always be verified with the Buckingham theorem, else it will not work. For problems involving many variables (e.g., in this book we consider problems involving up to nine variables), this process is tedious and performed manually by the user of this theorem.

5. Next, we set up the dimensional equations to form dimensionless groups. Taking one at a time,

   a. We first address a nondimensional pressure drop Δp in terms of the repeating parameters $\Pi_1 = \rho^a U^b d^c \Delta p \Rightarrow (\mathrm{ML}^{-3})^a \, (\mathrm{LT}^{-1})^b \, (\mathrm{L})^c \, (\mathrm{ML}^{-1}\mathrm{T}^{-2}) = \mathrm{M}^0\mathrm{L}^0\mathrm{T}^0.$

      We can write the solution to this in terms of linear independent equations as

      M: $0 = a + 0 + 0 + 1$
      L: $0 = -3a + b + c - 1.$
      T: $0 = 0 - b + c - 2$

      This can be solved by hand or using, again, using simple matrix algebra. We find

      $a = -1$
      $b = -2,$
      $c = 0$

      whence, $\Pi_1 = \dfrac{\Delta p}{\rho U^2}.$

   b. $\Pi_2 = \rho^d U^e d^f \, \mu \Rightarrow (\mathrm{ML}^{-3})^d \, (\mathrm{LT}^{-1})^e \, (\mathrm{L})^f \, (\mathrm{ML}^{-1}\mathrm{T}^{-1}) = \mathrm{M}^0\mathrm{L}^0\mathrm{T}^0.$

      We write

      M: $0 = d + 1$
      L: $0 = -3d + e + f - 1.$
      T: $0 = -e - 1$

We find

$$d = -1$$
$$e = -1,$$
$$f = -1$$

whence, $\Pi_2 = \dfrac{\mu}{\rho U d}$.

c. $\Pi_3 = \rho^g U^h d^i s \Rightarrow (ML^{-3})^g (LT^{-1})^h (L)^i (L) = M^0 L^0 T^0$.

We write

M: $0 = g$
L: $0 = -3g + h + i + 1$.
T: $0 = -h$

We find

$$g = 0$$
$$h = 0,$$
$$i = 0$$

whence, $\Pi_3 = \dfrac{s}{d}$.

d. $\Pi_4 = \rho^j U^k d^l \varepsilon \Rightarrow (ML^{-3})^j (LT^{-1})^k (L)^l (L) = M^0 L^0 T^0$.

We write

M: $0 = j$
L: $0 = -3j + k + l + 1$
T: $0 = -k$

We find

$$j = 0$$
$$k = 0,$$
$$l = -1$$

whence, $\Pi_4 = \dfrac{\varepsilon}{d}$.

6. Therefore, our function for the nondimensional parameters (termed here "Pi groups")
is

$$\frac{\Delta p}{\rho U^2} = F_1 \left(\frac{\mu}{\rho U d}, \frac{s}{d}, \frac{\varepsilon}{d} \right).$$

This concludes the process.

8.3   Use of a Large Number of Scaled Pressure Drop Measurements
to Close the Problem

In this section, we present an analysis wherein we again leverage experimental observations to achieve greater insight and help close the problem. The comments in this section are included in this chapter but are independent of the type of dimensional analysis used (whether it is Buckingham Pi theorem or Ipsen's method). Starting from the function in terms of nondimensional variables, we have

$$\frac{\Delta p}{\rho U^2} = F_1\left(\frac{\mu}{\rho U d}, \frac{s}{d}, \frac{\varepsilon}{d}\right).$$

We can simplify this by applying our rule ND2. Namely, the function of pressure drop is expected to vary in direct proportion to length s. Twice the length of pipe should lead to twice the pressure drop. For this reason, we should take care in our dimensional analysis to isolate the length to occur in but one nondimensional parameter, here s/d. We then apply rule ND2 and evaluate the known dependence as follows:

$$\frac{\Delta p}{\rho U^2} = \frac{s}{d} F_2\left(\frac{\mu}{\rho U d}, \frac{\varepsilon}{d}\right).$$

This is in fact correct and accurately predicts the data (White, 2016). To align with more conventional notation, we can now apply rule ND1 to manipulate the function into the following form:

$$\frac{\Delta p}{\frac{1}{2}\rho U^2} = \frac{s}{d} F_3\left(\frac{\rho U d}{\mu}, \frac{\varepsilon}{d}\right).$$

As per convention in the fluid mechanics community, we now define a factor f that is equal to the arbitrary function we derived from dimensional analysis as follows:

$$f = F_3\left(\frac{\rho U d}{\mu}, \frac{\varepsilon}{d}\right) = \text{friction factor for pipe flow.}$$

That is, the function F_4 is formally called the friction factor for pipe flow (or the Darcy–Weisbach friction factor) and is extremely useful in design of pipe systems. The shape of this three-variable function (describing the relation among three nondimensional variables) has been painstakingly mapped out for over 100 years of careful and tedious experiments. An example collection of such experiments is shown in figure 8.1. The figure shows a plot of friction factor f as a function of $Re_d = \rho U d/\mu$ and scaled roughness ε/d. In describing the importance of surface roughness to achieve dynamic similarity of the data of figure 8.1, Johann Nikuradse (1950) wrote:

Similitude requires that if mechanically similar flow is to take place in two pipes they must have a geometrically similar form and must have similar wall surfaces. The first requirement is met by the use of a circular section. The second requirement is satisfied by maintaining a constant ratio of the pipe radius r to the depth k of projections [*our ratio ε/d*]. It was essential, therefore, that the materials producing the roughness should be similar.

Figure 8.1 shows how friction factors for all pipe roughness collapse to a single curve of the form $f = 64/Re_d$ (a function predicted analytically by laminar flow theory) for Re_d less than about 2,300. The transition of f from a single straight line at low Re_d to multiple solutions at higher Re_d, each dependent on the value of ε/d is coincident with a transition from laminar to turbulent flow (cf. section 3.3) and the importance of inertial effects (including the effects of turbulent fluid flow interacting with roughness elements on wall). The shape of the three-variable function is still complex, but, in the plot of nondimensional numbers, we see that the data organizes nicely into a single curve per value of ε/d. Also, increases in ε/d result in a monotonic trend among the curves.

A large collection of the type of scaled data of figure 8.1 was obtained by Nikuradse and has continued to be been obtained and compiled by many over the last 100 years or so. These data have been analyzed and collated into a single plot known now as the Moody chart (after Lewis Ferry Moody). A detailed plot of the Moody chart can be found in any undergraduate text book on fluid mechanics. See, for example, figure 6.13 of White (2016) and or the large printed illustration of the Moody chart on the inside of the back cover of Fox et al. (2015). The Moody chart can be interpreted as a careful curve fit to a compilation of a large amount of pipe flow data that has been scaled in terms of the friction factor $f = f_4\left(\dfrac{\rho U d}{\mu}, \dfrac{\varepsilon}{d}\right)$. This famous chart is used by engineers and technicians to perform calculations of pressure drop in pipes of all shapes and materials.

The curves in the Moody chart are determined from experimental data. Fox et al. (2015) provide the following mathematical expression for a curve fit that is accurate to within a few percent error for flows in the turbulent region is:

$$\frac{1}{f^{0.5}} \approx -1.8\log_{10}\left[\frac{6.9}{Re_d} + \left(\frac{\varepsilon/d}{3.7}\right)^{1.11}\right].$$

This fit constitutes an approximate full closure of our pressure drop in a pipe problem. Without solving equations governing fluid flow, we have derived a useful relation among the key variables. That is, by analyzing experimental data in the context of dimensional analysis, we can determine the approximate underlying function. The fractional power of 1.11 reminds us that this is indeed a curve fit and not from basic principles.

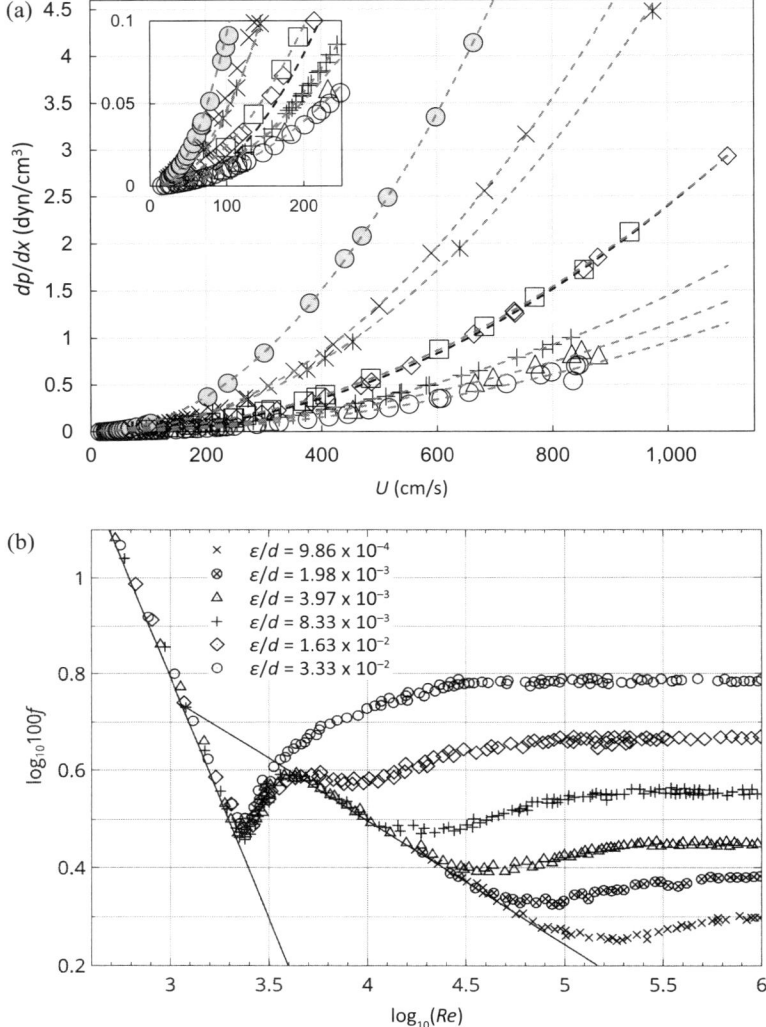

Figure 8.1
Plots relating pressure drop and velocity through a pipe. Top plot (a) shows measurements of dimensional pressure gradients along the pipe as a function of the area-averaged (aka bulk) velocity in the pipe. All are experimental data as reported by Nikuradse (1933, 1950). These data are only roughly half of the data reported by Nikuradse, and each series is roughly parabolic. However, the trends among the series are complex and nonmonotonic due to the effects of changing both viscosities and relative surface roughness. The bottom plot (b) shows the data collapse into a plot for nondimensional friction factor (the function F_4 of our analysis). Here a plot of log (base 10) of 100 multiplied by friction factor f as a function of log of $Re_d = \rho U d / \mu$ for various values of nondimensional roughness ε/d.

8.4   Buckingham Pi Theorem versus Ipsen's Method

In this book, we have chosen Ipsen's method over the better known and more commonly taught Buckingham Pi theorem (White, 2016; Yarin, 2012) for dimensional analysis. For readers who are users of the Pi theorem, we here propose that Ipsen's method has four advantages. First, Ipsen's method avoids the Pi theorem's tendency that multiple Pi groups contain the same repeating parameter. Instead, Ipsen's step-by-step method can more easily be used to specify the dimensional variables that appear within the various nondimensional parameters of the final expression. Second, for the Pi theorem, one must at times verify (manually) that the repeating parameters do not form a Pi group among themselves. The latter step is not needed for Ipsen's method. Third, unlike the Pi theorem, Ipsen's method requires no knowledge of linear algebra such as the meaning and application of determinants or matrix manipulation of any kind. Fourth, in my experience, Ipsen's step-by-step procedure method is more compatible and easily integrated with the rules of combining physical insight with dimensional analysis that I propose in this book, namely rules D1–D5 and ND1–ND6 proposed in chapters 5 and 6, respectively. For example, Ipsen's method can be initiated easily after applying rule D1, while the standard Pi theorem approach does not conventionally offer such flexibility in the initial hypothesis. The solution of the "overspecified problem" aspect associated with including a variable with a unique dimension (as per rule D2) follows from inspection of Ipsen method (i.e., we cannot eliminate a unique dimension using multiplications of or divisions by the other variables). In contrast, rule D2 is obscured within the conditions of the determinant calculated in the Pi theorem. Also, judicious choice of "chosen variables" of Ipsen's method as per section 5.2 can be used to achieve the goals of rules ND1 and ND2. That is, Ipsen's steps are, with some practice, more easily applied so as to guide which variables appear in which nondimensional parameter.

In my opinion, Ipsen's method is more intuitive and insightful than Buckingham Pi theorem. Ipsen leads us inexorably to the correct number of nondimensional groups every time. The slight "time saving" of programming Buckingham Pi steps (e.g., matrix manipulation) is not worth the cost distancing oneself from the variables and analysis.

8.5   Summary

- For completeness, we summarized the Buckingham Pi theorem as a comparison method for dimensional analysis. The Buckingham Pi theorem leverages some concepts of linear algebra to aid identification of nondimensional parameters called Pi groups.
- Several steps of the Buckingham Pi theorem lend themselves to programming via linear algebra, but the method also requires manual and sometimes tedious verification

that the so-called repeating parameters do not form a Pi group among themselves (otherwise the method will fail).

- As an example application of the Pi theorem (and an very important example of dimensional analysis in general), we analyzed pressure drop of fluid flow through pipes with finite surface roughness.

- After taking dimensional analysis as far as it would go, we applied rule ND2 and used physical intuition to posit that the pressure drop in the pipe should scale directly proportional with its length s. As per rule ND2, we isolated the nondimensional variable involving pipe length, s/d, and then explicitly evaluated this dependence by taking it as a prefactor to a new unknown function involving the rest of the parameters.

- The Moody chart visualizes the three-dimensional dependence among nondimensional pressure drop (termed the friction factor), Reynolds number, and relative surface roughness. Interestingly, the first of these depends mostly on the third for large values of the second and third.

- We discussed several advantages of Ipsen's method over the more commonly taught Buckingham Pi theorem. Perhaps the most important advantage of Ipsen is that it is easily compatible with the rules proposed in this book (and summarized in section 11.2) around integration of dimensional analysis with physical intuition and experimental observations.

Problems

8.1. Apply the Buckingham Pi theorem to relate the frequency of oscillation of a guitar string ω to its tension T, length L, and mass m.

8.2. Repeat the surface wave dynamics problem 6.1, but this time using the Buckingham Pi theorem to derive the nondimensional variables.

8.3. Consider again the pressure drop in a pipe as described in this chapter. Repeat the Buckingham Pi theorem approach on the function

$$\Delta p = \quad f_1(U, \quad d, \quad s, \quad \varepsilon, \quad \rho, \quad \mu).$$
$$ML^{-1}T^{-2} \quad LT^{-1} \quad L \quad L \quad L \quad ML^{-3} \quad ML^{-1}T^{-1}$$

However, purposely make the poor choice of the following three repeating parameters: U, d, and s. (These repeating parameters form a nondimensional group among themselves, d/s.) What goes wrong?

9

Leveraging of Model Data to Build and Understand Prototypes

At first we had taken up the problem merely as a matter of sport, but now it was apparent that if we were to make much progress it would be necessary to get better tables from which to make our calculations. In September we set up a small wind tunnel.... We made thousands of measurements of the lift, and the ratio of the lift to the drift.

—Orville Wright (1953)

In this chapter, we summarize some principles of leveraging experimental data from models to the design and study of prototypes. A prototype refers to an early, full-scale working model of a machine or system. For example, a prototype might be a fully functional airplane or a full-scale vacuum cleaner robot. The word "model" refers to a construct or assembly used to predict the behavior or performance of the prototype, and thereby save money. For example, a 1/20th scale model might be easier and cheaper to fabricate and study experimentally than the prototype. Consider for example a small model of an airplane fabricated from a polymer using 3D printing and placed in a wind tunnel.

9.1 Geometric, Kinematic, and Dynamic Similarity

The three mathematical expressions relating measurements and conditions of models to those of a prototype are called similarity rules. We must identify relations that relate measurements of the models to predicted corresponding behavior the prototype. These relations are known as *scaling laws* between model and prototype. Consider this question:

Question 9.1:   What aspects of the model and prototype should be similar for us to make predictions on prototype based on experiments with the model?
    Basically, there are three similarity conditions that must be met:

(1) *Geometric similarity:*   This is the obvious requirement that the model and prototype be the same shape. So, for example, all of the ratios of lengths of respective features and all of the angles are the same between the model and the geometrically similar prototype. See also section 4.1.

(2) *Kinematic similarity:* This requires the velocities (including velocities of fluid interacting with solid bodies) at corresponding/homologous points in the model versus prototype be in the same direction and that their magnitudes be related by a constant scale factor. For example, the streamline patterns will also be related by a constant scale factor. Note that this requires geometric similarity. For example, in fluid flow, the boundaries of a body form partly form the shape of streamlines. Kinematic similarity also requires that the same flow regime be true for both model and prototype. For example, if compressibility, surface tension, or free surface effects are all absent from the prototype flow, they must be avoided in the model flow.

(3) *Dynamic similarity:* This requires that, in the model and prototype, identical types of forces at corresponding points be parallel and be related by a constant scale factor throughout any objects or flow fields. This is the most restrictive requirement, since dynamic similarity requires both geometric and kinematic similarity.

Dynamic similarity ensures that model and prototype (indeed any two versions of the same phenomena with different absolute scales) will be relatable. This leads us to the answer of our earlier question.

Answer to question 9.1: We need the condition of dynamic similarity. When achieved, model data can be used to quantitatively predict prototype conditions including prototype kinematics (e.g., position and velocity) and dynamics (forces).

> Comment: Note that, in general, the scale factors for geometric (lengths), kinematic (velocities), and dynamic (forces) are, in general, different from one another. For example, a model sphere may be 1/5th the diameter of a prototype sphere, while the drag force on the model sphere may be 1/25th that of the prototype.

9.2   Experimental Design and Interpretation: To Match All Variables, Match All but One

We here describe the mechanics of deriving similarity laws. Consistent with our use of Ipsen's method (vs. Pi theorem), we will use N_i to indicate the *i*th nondimensional group (vs. Buckingham's Π_i notation).

Question 9.2: How do we ensure dynamic similarity between model and prototype flows?

An answer to question 9.2: For dynamic similarity, each value of a nondimensional parameter of the model must match the respective value for that same parameter evaluated for the prototype. For *m* Pi groups (e.g., derived using Ipsen's method) associated with the essential physics of the problem, we have

$$N_{1, \text{model}} = N_{1, \text{prototype}}$$

$$N_{2, \text{model}} = N_{2, \text{prototype}}$$

. . .

$$N_{m, \text{model}} = N_{m, \text{prototype}}.$$

For example, for the submarine drag from section 3.4, we would ensure that

$$Re_{\text{model}} = Re_{\text{prototype}}$$

$$C_{D, \text{model}} = C_{D, \text{prototype}}.$$

Now, our hypothesized functional relationship implies that we must match all but one nondimensional parameter. Consider the functional relation for either prototype or model:

$$N_m = F(N_1, N_2, N_3, \dots N_{m-1}).$$

If our hypothesis is correct, then knowledge of $N_1, N_2, N_3, \dots N_{m-1}$ determines N_m. Hence, we can say that if $(N_1, N_2, N_3, \dots N_{m-1})$ for model is known then N_m for model is known. If we match $(N_1, N_2, N_3, \dots N_{m-1})$ for model to prototype, then N_m of prototype will also be matched, since

$$N_{m, \text{prototype}} = F(N_{1, \text{prototype}}, N_{2, \text{prototype}}, N_{3, \text{prototype}}, \dots N_{m-1, \text{prototype}}).$$

This leads us to an improved answer to question 9.2.

A better answer to question 9.2: Match the $m - 1$ independent nondimensional variables in the model with the respective nondimensional groups of the prototype. If the hypothesized function is correct, this ensures the dependent variable, the mth group, will also be matched. In this notation

- Match $(N_1, N_2, N_3, \dots N_{m-1})$ between model and prototype
- If so, then for the dependent variable will also be matched: $N_{m, \text{model}} = N_{m, \text{prototype}}.$

Note that similarity can be defined as the condition that each of the nondimensional "natural variables" of the model be the same as each respective nondimensional parameter for the prototype.

9.3   Examples of Model and Prototype Studies

We will here discuss a few example applications of dimensional analysis wherein measurements on a model are used to think about and design a prototype. The relations between them are called similarity laws.

9.3.1 Submarine Prototype Scaling

Consider again the submarine from sections 4.2–4.3 and 5.3. Imagine we use three-dimensional printing to create a 1/10th scale model of the prototype. We want to use a force measurement sensor and a wind (not water) tunnel blowing air at standard conditions to estimate the drag versus velocity curve for the prototype, which will travel at 0.5 m/s. At what air velocity do we run the wind tunnel? How do we relate the model's drag to the prototype's drag?

We proceed as follows. Our known information is as follows (where subscripts m and p respectively denote model and prototype):

Model	Prototype
$10b_m = b_p$	
$v_m = 1.51e - 5\ \mathrm{m^2/s(air)}$	$v_p = 1.01e - 6\ \mathrm{m^2/s(water)}$
$V_m = ?$	$V_p = 0.5\ \mathrm{m/s}$

As we saw in chapters 4 and 5, the relevant function relating the nondimensional terms is

$$\frac{F}{\rho V^2 b^2} = F\left(\frac{Vb}{v}\right). \tag{9.1}$$

For convenience, we have formulated the Reynolds number in terms of kinematic viscosity defined as $v = \mu/\rho$. For dynamic similarity, we recognize that, if we match Reynolds number, then the drag coefficient will also be matched (cf. the "better answer" to question 9.2 in section 9.2). Hence, we write the similarity laws as follows.

If

$$\left.\frac{Vb}{v}\right|_m = \left.\frac{Vb}{v}\right|_p , \tag{9.2}$$

Then

$$\left.\frac{F}{\rho V^2 b^2}\right|_m = \left.\frac{F}{\rho V^2 b^2}\right|_p . \tag{9.3}$$

From equation (9.2),

$$V_m = \underbrace{\frac{b_p}{b_m}\frac{v_m}{v_p}}_{\substack{\text{scaling from our choices} \\ \text{of model scale and fluid}}} V_p = 10 \cdot \frac{1.51e-5}{1.01e-6} V_p = 149.5\ V_p = 75\ \mathrm{m/s}$$

That is, our model is so small (1/10th the length of the prototype) and the kinematic viscosity of water is so much smaller than that of air ($v_m/v_p = 14.95$) that we must run the model in the wind tunnel at a velocity 149.5 times higher than the prototype velocity through water. Note that if our wind tunnel cannot run at this high velocity (roughly 270 km/h or 168 mph for this example), then we must revisit our choice of using air or consider building a larger scale model. For example, perhaps we use a 3D printer to create four sections and then assemble these.

If the wind tunnel can handle the velocity of our original estimate, then we can relate the drag from similarity law equation (9.3):

$$F_p = \underbrace{\frac{\rho_p}{\rho_m}\frac{V_p^2}{V_m^2}\frac{b_p^2}{b_m^2}}_{\substack{\text{scaling from our choices} \\ \text{and requirement of } Re_m = Re_p}} \qquad F_p = \frac{998}{1.2}\cdot\frac{1}{149.5^2}\cdot 10^2 \cdot F_m = 3.72\, F_m.$$

So the magnitude of the drag forces measured in the model should each be multiplied by about 3.72 to predict the corresponding prototype forces.

9.3.2   Boat Drag and How Scale Model Studies Are Not Always Possible

We wish to estimate the drag caused by the water on the boat (the hull drag) as a function of velocity for a planned full-scale prototype. Hull drag is a complex function of the speed of the boat relative to the water, the orientation of the boat (e.g., a heeling sailboat), and the conditions of the water surface (e.g., the size and shape of waves). The image of figure 9.1 helps capture some of the complexities in the problem. We here pursue a fairly simplified versions of these dynamics.

We here start with a simple problem: the case of an upright boat traveling steadily in otherwise calm water. We build a geometrically similar model and fabricate this model with the correct volume-averaged density so that (when it is at rest) it floats on the water at the correct waterline (so the waterline position of model versus prototype will be geometrically similar). Our model is positioned in a water tank and towed through the water. We make measurements on the model and try to extrapolate these to the planned prototype.

Hull drag is due to both shear forces of viscous drag and to pressure forces on the hull. The pressure forces are due to the coupling of the water velocities and the deformed shape of the free surface. For example, even in calm water, the motion of a boat itself disturbs the water and creates waves. In essence, you can think of the motion of a boat as being resisted by a wave of its own making.

Consider geometrically similar boats of length L. We postulate the following key variables:

Figure 9.1
A sloop with a reefed mainsail on a starboard tack. The boat's hull interacts with both ocean waves and waves of its own making. Source: http://www.pexels.com.

Variable	Definition	Physical quantity
F_D	drag force on boat	MLT^{-2}
U	velocity of boat	LT^{-1}
L	length of boat at water line	L
g	gravitational acceleration	LT^{-2}
ρ	density of fluid	ML^3
μ	dynamic viscosity of fluid	$ML^{-1}T^{-1}$

We hypothesize the following function for drag:

$$F_D = f(U, L, g, \rho, \mu).$$

Here, as per rule D5, we have invoked geometric similarity (for the submerged part of the boat in contact with water) and so use a single length scale L to describe the geometry.

We have six variables and three physical quantities. We apply Ipsen's method and derive the following:

$$\frac{F_D}{\rho U^2 L^2} = F\left(\frac{\rho UL}{\mu}, \frac{U^2}{Lg} \right).$$

We can invoke rule ND1 to divide the left-hand side by 0.5 and write as

$$C_D = G(Re, Fr^2),$$

where we again see the trade-off between the kinetic and potential energy (here, of wave motion) as characterized by Froude number, Fr.

Now, for dynamic similarity we require the following similarity laws between model and prototype:

$$\left.\frac{\rho UL}{\mu}\right|_m = \left.\frac{\rho UL}{\mu}\right|_p$$

$$\left.\frac{U^2}{Lg}\right|_m = \left.\frac{U^2}{Lg}\right|_p.$$

So that, given our hypothesized function, we will have

$$\left.\frac{F_D}{\rho U^2 L^2}\right|_m = \left.\frac{F_D}{\rho U^2 L^2}\right|_p.$$

Now, consider a boat model constructed to 1/20th scale and used to predict the hull drag of a prototype to operate at 10 m/s. At what velocity should the model test be run? Is water a possible liquid for the test? First, we review our information:

Model	Prototype
$20L_m = L_p$	
$U_m = ?$	$U_p = 10\,\text{m/s}$
$v_m = ?$	$v_p = 1.01\text{e} - 6\,\text{m}^2/\text{s (water)}$

Matching model and prototype Froude numbers, we have this similarity law

$$\left.\frac{U^2}{Lg}\right|_m = \left.\frac{U^2}{Lg}\right|_p \Rightarrow \frac{U_m^2}{U_p^2} = \frac{L_m}{L_p}\frac{g_m}{g_p}.$$

Both model and prototype will operate on Earth's surface, so $g_m/g_p = 1$ and

$$\frac{U_m}{U_p} = \left(\frac{L_m}{L_p}\right)^{1/2} = \left(\frac{1}{20}\right)^{1/2} = 0.224 \Rightarrow U_m = 0.224\, U_p = 2.24 \text{ m/s}.$$

From the Reynolds number similarity

$$\left.\frac{UL}{\nu}\right|_m = \left.\frac{UL}{\nu}\right|_p \Rightarrow \frac{\nu_m}{\nu_p} = \underbrace{\frac{U_m}{U_p}}_{\substack{\text{demanded by}\\ Fr\text{ similarity}}} \underbrace{\frac{L_m}{L_p}}_{\substack{\text{from our}\\ \text{choice of scale}}} = 0.224\,\frac{1}{20} = 0.0112.$$

The model test liquid cannot be water! In fact, it must have a kinematic viscosity of about 1% that of water. If we glance through table A.1 of the appendix and similar data, we quickly see that no such available liquid exists (aside perhaps for some exotic cryogenic superfluid).

We come upon a great difficulty in predicting prototype designs for ships: It is very difficult to match both Re and Fr simultaneously. See Fox et al. (1985) for further discussion. The reason for this becomes clear if we consider the difference in dependence between velocity and length for the two relevant scaling laws. Assuming water is used for both model and prototype (and both will operate in Earth's gravity), then we can rewrite the dynamic similarity conditions in dimensional form as follows:

$$\left. UL \right|_m = UL_p \quad \text{vs.} \quad \left.\frac{U^2}{L}\right|_m = \left.\frac{U^2}{L}\right|_p.$$

The first relation implies that the smaller the model, the higher should be its velocity to match Re. Conversely, the second relation implies that the smaller the model, the lower its velocity should be to match Fr. It is not easily achievable (and often impossible) to achieve dynamic similarity with a small-scale model.

This simple comparison helps explain why several countries have, at great expense, built giant "model basin" tanks to test near-full-scale "models" of planned prototype boats. One such tank is the historic David Taylor Model Basin in Bethesda, Maryland. This facility included a test tank 2,968 ft long by 21 ft wide and depths of up to 16 ft and housed in a 3,200 ft long building. It was designed to accommodate model ships up to 40 ft in length and tow them up at speeds of up to 50 knots (ASME, 1998).

9.4 Summary

- To be useful, model experiments should be geometrically, kinematically, and dynamically similar to prototype experiments. Dynamic similarity requires geometric and kinematic similarity.

- Dynamic similarity implies that all the important/relevant nondimensional parameters of the model are equal to the respective nondimensional parameters of the prototype.
- To ensure the latter condition, we need to match all but one of the nondimensional parameters in the problem. Doing so ensures that the last one will necessarily be matched.
- For some problems, matching all nondimensional parameters may not be feasible or even possible.
- In our example of boat drag, we concluded that matching both Froude and Reynolds numbers is very difficult because of the respective inverse scalings these have with respect to length and velocity.

Problems

9.1. A prototype submarine is to travel at 1 m/s in the ocean. You wish to construct a 1/50 scale model that will be tested in fresh water. What should be the speed for model test? What is the ratio of prototype to model drag force? Hint: The density of sea water is 1,027 kg/m^3, while that of fresh water is 998 kg/m^3. The viscosities are approximately the same ($1.0e^{-3}$ $kg \cdot m \cdot s^{-1}$).

9.2. You observe experimentally that water flow through a circular tube with diameter $d = 2$ cm becomes turbulent at a bulk velocity of 11.5 cm/s. For airflow in the same tube, what is the volume flow rate in m^3/s that causes transition to turbulence?

9.3. Non-Newtonian fluids are fluids that do not follow Newton's proportional relation between rate of strain and shear stress. One relevant nondimensional number is the Bingham number, Bm, which is used to characterize relative strength of shear and elastic stress and has the form

$$Bm = \frac{\tau_y d}{\mu V}.$$

Here τ_y is the yield stress of the non-Newtonian fluid, and μ is dynamic viscosity. V and d are characteristic velocity and length scales, respectively. You wish to study the dynamics of (including forces on) smooth spheres moving through such a fluid using a model sphere with a diameter that is 1/10th the diameter of the prototype sphere. Both model and prototype will be operated in the exact same fluid. What is the correct model-to-prototype velocity ratio to match the model versus prototype Bm? Can you easily match both Bm and Reynolds numbers?

10

Small Changes in Geometry Can Have Significant Effects: The Effect of Roughness on Drag

> Eiffel found that the resistance to a sphere moving through air changes its character somewhat suddenly at a certain velocity. The consideration of viscosity shows that the critical velocity is inversely proportional to the diameter of the sphere.
>
> —Lord Rayleigh (1915)

There are experimental data available for many types of physical phenomena. However, the problem of interest to you may not have been studied exactly. It may be very tempting to roughly approximate predictions for one geometry given data for a similar geometry (indeed we saw in chapter 7 a scaling that approximately equated the shape of a mosquito and airliner). However, it is not always the case that the behavior of one geometry can be used (i.e., "stretched") to predict behavior of another. In nonlinear phenomena, a tiny change can have a big effect. Seemingly small differences in geometry can at times have significant effects. We have already seen one example of this in chapter 8, where we discussed the significant importance of pipe inner surface roughness on pressure drop. In this chapter, we discuss in some detail a similar problem: the significant effect that relatively small roughness elements can have on the drag of immersed bodies. We also here discuss how drag scaling laws obtained for one shape might be used to roughly predict the drag experienced by different shapes.

10.1   Roughness and the Drag of Spheres

We first discuss an example where even slight changes in geometry can have a big effect, making it difficult to approximate the behavior of one body given data for another. In section 3.4.2, we discussed drag coefficients of geometrically similar, smooth spheres. How well do these results apply to a sphere with a roughened surface? Not very well, it turns out. One reason for this is that even the 0.1% relative roughness can have pronounced effects on the near-wall flow of fluid (the boundary layer).

The physical situation and influence of sphere roughness is depicted schematically in figure 10.1. Roughness elements have a strong effect on the regions of the fluid flow very

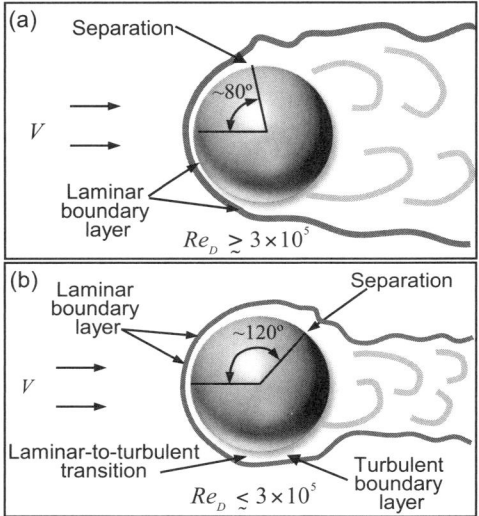

Figure 10.1
Laminar-to-turbulent transition of the boundary layer can significantly decrease drag on a sphere. Shown are schematics of flow around a sphere for Re_d values less and greater than about 3×10^5. For low Re_d, the boundary layer remains laminar and this relatively low inertia flow separates early (at low separation angle of about 80°) from the sphere, causing a large wake (a low pressure region). For rougher spheres and higher Re_d, the boundary layer transitions to turbulent. The turbulent flow separates at a higher angle (near 120°), resulting in a smaller wake. Roughness elements can cause early transition to turbulence, and so a small change in geometry (relative height of small roughness elements) can have a large effect on visible behavior.

near the surface known as the boundary layers (see section 3.3). For Re_d greater than about 10, the boundary layers are very thin regions near the surface where the balances among viscous forces, pressure forces, and inertia are important. In this region, the fluid imparts a high shear force on the body. At the very front of the sphere, the flow is decelerated and there is accordingly a region of high pressure (dynamic pressure effect), while the wake is a region of relatively low pressure. Higher inertia, turbulent boundary layers do a better job of wrapping around the sphere and persist further into the low-pressure region. Roughness elements favor turbulent boundary layers and hence turbulent boundary layers are able to wrap around the sphere more efficiently, resulting in a smaller wake and less pressure-based drag. The situation may at first seem counterintuitive (since turbulent boundary layers have greater shear stress than laminar boundary layers). However, as we shall discuss further in section 10.3, the effects of pressure forces on drag are here more important than shear stress forces, and so overall drag is reduced by the smaller wake of the turbulent boundary layers. See similar discussions in Incropera et al. (2013) and White (2016).

A correct application of dimensional analysis would proceed as follows. We first hypothesize the following parameters as controlling the drag on spheres with finite roughness:

Variable	Definition	Physical quantity
F_D	drag on a sphere (dependent variable)	MLT^{-2}
V	external flow velocity	LT^{-1}
d	diameter of sphere	L
ε	roughness element length (e.g., standard deviation height of surface features)	L
ρ	density of fluid	ML^3
μ	dynamic viscosity of fluid	$ML^{-1}T^{-1}$

Such a list implies the following hypothesis for the functional dependence:

$$F_D = f_1(V, \ d, \ \varepsilon, \ \rho, \ \mu).$$
$$MLT^{-2} \quad LT^{-1} \ \ L \ \ L \ \ ML^{-3} \ ML^{-1}T^{-1}$$

We can then apply dimensional analysis to derive

$$\frac{F_D}{\rho U^2} = G\left(\frac{\mu}{\rho U d}, \frac{\varepsilon}{d} \right).$$

We invoke rule ND1 to reformulate this in terms of a more conventional nomenclature as follows:

$$C_D = F\left(Re, \frac{\varepsilon}{d} \right),$$

where

$$Re = \frac{\rho V d}{\mu} = \text{Reynolds number}$$

$$C_D = \frac{F_D}{\frac{1}{2}\rho V^2 \left(\frac{\pi d^2}{4} \right)} = \text{drag coefficient of diameter } d.$$

Figure 10.2 shows a plot of a wide range of experimental data for the drag of rough and smooth spheres in terms of the nondimensional parameters of drag coefficient C_D and Reynolds number Re. As shown by these measurements by Achenbach (1972, 1974), spheres with turbulent boundary layers have smaller wakes than laminar spheres and so significantly less drag. Further, spheres with rough surfaces promote earlier transition of laminar to turbulent boundary layers.

The drag force measurements of figure 10.2 for smooth spheres are from Achenbach (1972), and the data for spheres of various nondimensional roughness ε/d (where ε is the characteristic length scale of the roughness elements) are from Achenbach (1974). The

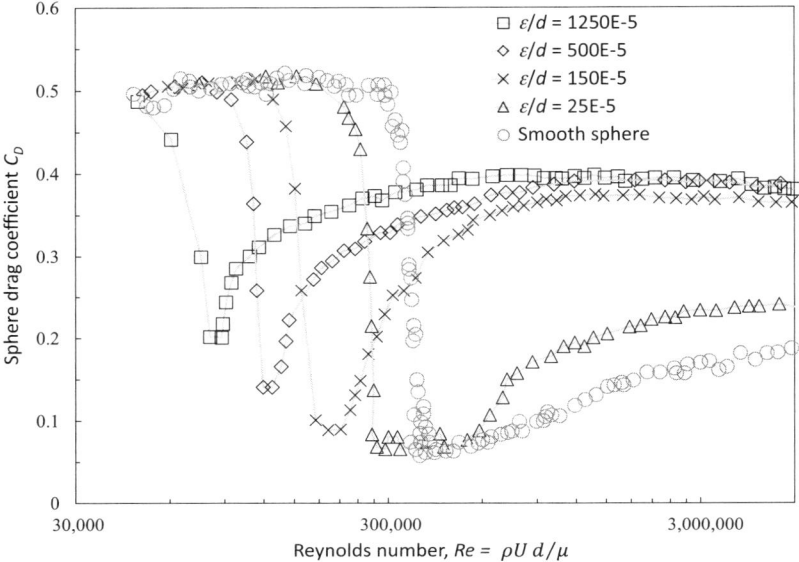

Figure 10.2
Drag coefficient of smooth and rough spheres as a function of Re_d and characteristic roughness. The smooth sphere data was digitized directly from Achenbach (1972), and the rough sphere drag data from Achenbach (1974). The relative scale of surface roughness can have an immense effect on drag, particularly at moderate to high Re_d.

sharp drops in drag coefficient for figure 10.2 (often termed the "crisis" region in these drag plots) can be roughly interpreted as regions where the boundary layers become turbulent and so suddenly begin to more efficiently wrap around the sphere, creating a smaller wake. Achenbach notes the critical Reynolds number at which this drop occurs decreases monotonically with Re. This effect of roughness can have profound effects on drag coefficient—even though it is a small (and sometimes microscopic) change in geometry.

See Blevins (1984) and Choi (2006) for discussions of the drag coefficient of (rough) golf balls and experiments showing how golf balls can fly further than an equally sized and weighted smooth balls.

The preceding discussions point out that the shape and relative (to diameter) scale of roughness elements can have a significant effect on drag. Hence, predicting drag of a rough sphere from smooth sphere data is not accurate. The general lesson is that fluid flow problems of moderate and high Reynolds numbers are sufficiently complex and nonlinear that even tiny changes in the geometry (here the size of small roughness elements) can have a strong effect on macroscopic behavior such as the drag.

10.2   Pressure Drop in Smooth versus Rough Pipes

As a brief, second example, consider the analysis for pressure drop in a pipe presented in section 8.3. From the experimental data of Nikuradse we see that at $Re_d = \rho U d / \mu = 10^5$, roughness elements of characteristic scaled length of $\varepsilon/d = 0.1\%$ can raise the friction factor from about 0.0085 (for a smooth pipe) to 0.02. That is, a 0.1% change in the scale of the surface roughness of the interior of the pipe alone can increase absolute pressure drop (e.g., measured in kPa) in the pipe by a factor of 2.4.

10.3   Estimating Drag from Different Shapes

Despite the strong sensitivity of surface roughness, we may nevertheless wish to obtain even rough estimates of the drag of one body given data from a similarly shaped body. Taking a lesson from the first part of this chapter, we first might strive to find data with roughly the same relative roughness ratio ε/d, where d is the characteristic macroscopic dimension for the body of interest. However, which *macroscopic* geometric scale should we choose to compare two differently shaped bodies? Consider first a generalized drag coefficient and Reynolds number, defined as follows:

$$Re = \frac{\rho V \ell}{\mu} = \text{Reynolds number}$$

$$C_D = \frac{F}{\frac{1}{2}\rho V^2 A} = \text{drag coefficient}$$

V is the free-stream (approaching) velocity. The length scale ℓ in Re is typically the chord length or body length parallel to the stream. For cylinders, spheres, and disks, ℓ is typically a diameter.

White (2016) provides some advice on how to roughly estimate drag data of one geometric shape given drag data for another slightly different shape. An important choice is the area A variable in the drag coefficient. The following are good choices for three typical cases:

1. Frontal area A (the area of the body projected into the direction of approach free stream): This area is most appropriate for nonstreamlined bodies—that is, so-called "bluff" bodies (think of stubby, thick bodies like a falling meteor or a sphere).

2. Planform area A for streamlined bodies and/or very wide, flat bodies: For such bodies, this is the area of the body projected in a direction perpendicular to the free stream and

often the maximum projected area. For example, planform area is the area of the shadow of streamlined airplane in level flight at high noon.

3. Wetted area A for surface ships and barges: This is the (curved) surface area in contact with the water during motion.

A judicious choice of the characteristic area (using the preceding guidelines) for various drag coefficients helps us approximate drag coefficients for which we may not have direct data. For example, for a smooth, streamlined body (a specific torpedo shape or airfoil), we may predict absolute drag using drag coefficients calculated based the planform area A, since the body's drag is a strong function of this area.

As another example, we might estimate the drag on a human body falling (in a horizontal, spread-eagle position similar to a skydiver as discussed in section 4.4.3) using the frontal area of a similar shape. If there is no similar shape available, we might use the closest available. For example, we might use the data for a finite cylinder with the air flow in a direction perpendicular to the cylinder's axis. We would use the $C_D(A)$ vs. Re data with length scale ℓ for Re as the "diameter" of the cylinder of the person's torso and A equal to projected area of the person. Loth (2008) also discusses the approximation of drag coefficients of irregular shapes.

10.4 Example of Incorrect Dimensional Analysis: Neglecting a Parameter

Consider again the dimensional analysis of the section 10.1 for the drag on spheres. As mentioned earlier, the problem of drag coefficients on bodies with a finite surface roughness is complex because seemingly slight differences in the geometry (the relative height of small roughness elements) can have a profound effect on drag, particularly at moderate and high values of Reynolds number. Achenbach's (1974) data was obtained with a 0.2 m diameter sphere, so the smallest controlled roughness element length he reported was on the order of 50 microns. Consider that, at $Re = 300,000$, such 50 micron scale roughness elements can reduce the drag coefficient (and hence the absolute drag force on a sphere of equal diameter) by about a factor of 10 (relative to a smooth sphere). As discussed in section 10.1 and shown in figure 10.1, this extreme sensitivity to minute (even microscopic) differences in geometry is due to the effect of the surface roughness on the boundary layers of the flow around the sphere. Small roughness elements can cause the boundary layer to become turbulent so that it has greater ability to wrap around the sphere before it separates, resulting in a smaller wake.

Given this tiny "nearly hidden" effect of surface roughness, consider now an incorrect (but understandable) application of dimensional analysis wherein we neglect the parameter describing the characteristic height of roughness elements in the analysis. This would lead to the following table of parameters:

Variable	Definition	Physical quantity
F_D	drag on a sphere (dependent variable)	MLT^{-2}
V	external flow velocity	LT^{-1}
d	diameter of sphere	L
ρ	density of fluid	ML^3
μ	dynamic viscosity of fluid	$ML^{-1}T^{-1}$

This list is consistent with the following hypothesis for the functional dependence:

$$F_D = f_1(V, \ d, \ \rho, \ \mu),$$
$$ML^{-1}T^{-2} \quad LT^{-1} \quad L \quad ML^{-3} \ ML^{-1}T^{-1}$$

and the following incorrect result for the problem.

$$C_D = F(Re).$$

Dimensional analysis alone does not point out any problems. We are simply missing a key parameter: the length scale of roughness elements. The problem would arise when we begin to obtain and analyze quantitative experimental measurements. Figure 10.3 depicts the type and extent of the problems that would arise. This figure depicts the large variation of C_D and Re implied by the measurements of Achenbach (1972, 1974) as shown in figure 10.2. The gray area shown in the plot are the possible locations of drag coefficient measurements given values of Re. Hence, for the current example, we measure drag forces across with spheres of varying diameters, in flows with varying velocities, and in fluid of varying properties, but we would ignore the roughness of the surfaces of our test spheres. The data would easily be scattered throughout the entire grey area in the plot. The two nondimensional parameters, C_D and Re, would help us identify a "neighborhood" of solutions, but never a single valued functional relationship for drag coefficient given Re. For example, we would find that a single value of Reynolds number can result in a wide range of drag coefficients (as depicted by the vertical column of circles). We would reasonably (and correctly) conclude that our "independent variable" (Re) is not sufficient to specify drag force. Clearly, we have not well specified the problem as the dependent parameter varies widely for equal values of the independent variable. This would be the strongest evidence that our list of parameters was incomplete, and this might motivate us to look for other "hidden" variables that we are not yet controlling.

Recall from figure 10.2 that the local minimum in C_D associated with the drag "crises" explained earlier. This local minimum implies that there are at least two values of C_D for each value of the independent parameter Re, even when we account for roughness. This double value of drag is associated with regimes where shear and pressure drag respectively dominate in regions of low and high Re (cf. section 10.2). In the current example, where we ignore roughness, we would see not two but a wide range of Re values for each

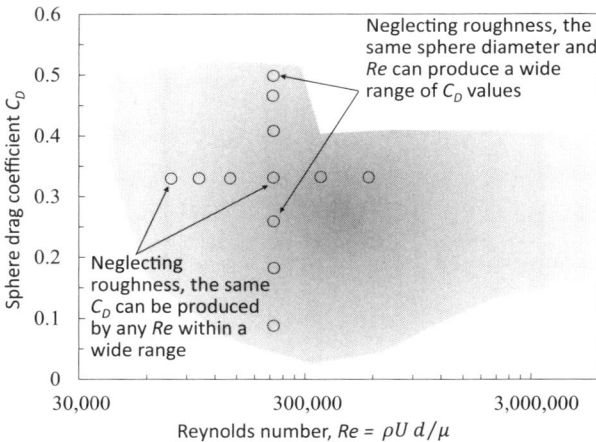

Figure 10.3
Example of the type of ambiguity that would result from mistakenly excluding the parameter describing surface roughness height in the dimensional analysis associated with the data of figure 10.2. Most important, one value of Re would yield many and widely different values of drag coefficient C_D (as depicted by the vertical column of circles). Also, a wide range of Re could easily yield the same drag coefficient (horizontal row of circles).

value of C_D (particularly in the range below about $Re = 300{,}000$). Again, this ambiguity would be evidence that we are missing an important variable.

Another ambiguity would arise in comparing data between two laboratories. Imagine that one lab uses exclusive one diameter and one roughness (e.g., while varying velocity). A second lab uses a second diameter and a second roughness. Even if the labs exactly match Re, their measured drag coefficients will never match unless they both (by some coincidence) used the same respective values of ε/d.

In brief, missing an important parameter leads to ambiguities that become clear only after experiments are performed.

10.5 Summary

- We found that approximating quantities in one geometry using nondimensionalized experimental data for an approximate geometry is not always possible. As an example of this, we noted how the drag coefficient function versus Reynolds number for rough spheres (in turbulent flow) is very different from that of smooth spheres. We saw a similar disparity for pressure drop in smooth versus rough pipes in chapter 8.

- We discuss some rules of thumb in making rough estimates of drag on a macroscopic shape given drag coefficient data for similar geometries (e.g., given approximately same surface roughness). For bluff bodies, we scale drag and lift coefficients with projected area (projected in flow direction). For streamline bodies, we use planform area. For streamlined boats, we use wetted area.

• Last, we discussed an example of the type of ambiguities that could arise from missing an important variable in dimensional analysis: neglecting roughness element heights in the drag data for spheres. The ambiguities would become increasingly clear as we obtain and analyze experimental data. Missing an important variable would also preclude meaningful lab-to-lab comparisons.

Problems

10.1. E. Achenbach's (1974) rough sphere data was obtained from experiments with a 0.2 m diameter sphere. He imparted roughness by, for example, abrading the surfaces with emery paper. What characteristic roughness height (in microns) would you require to reproduce his data for relative roughness of $\varepsilon/d = 25E-5$ using a 2 cm diameter sphere?

10.2. The Robins–Magnus effect refers to the transverse force experienced by a spinning sphere or cylinder in a cross flow. Think of the curve flight of a spinning ball in baseball or golf. A golf ball (4.27 cm diameter) can spin as fast as roughly 8,000 rpm as it travels through the air at 20 m/s.

   a. You wish to reproduce these conditions in a wind tunnel with a maximum wind velocity of 20 m/s. What model ball diameter do you need? How fast should the model golf ball spin?

   b. If the prototype (standard) golf ball has dimple depths of about 0.25 mm, what will be the dimple depth of the model golf ball?

10.3. Approximate the shape of your body tucked into a "ball" as a sphere. Estimate your terminal velocity in this position. For the function relating drag coefficient versus Reynolds number, use the experimental data given by Achenbach for rough spheres (see figure 10.2). Note that Achenbach defines his drag coefficient in the usual manner with a factor of 0.5 in the denominator.

11

The Riabouchinsky–Rayleigh Paradox and the Rule of Relevance

In this chapter, we discuss a discussion between Lord Rayleigh and Dimitri Riabouchin-sky that occurred shortly after publication of Rayleigh's (1915) paper on dimensional analysis. The published discussion between these two important figures seemed to point to a paradox in dimensional analysis—namely, that the incorporation of physical insight seemed, in at least one case, to increase the complexity of the problem. This chapter describes the resolution of this issue and then proposes a generalized rule that we can apply to all dimensional analyses. I call this the "rule of relevance." At the end of this chapter, we summarize the twelve rules that we have proposed in this book for combining physical intuition and experimental observations with dimensional analysis.

11.1 The Riabouchinsky–Rayleigh Paradox

In his seminal paper on dimensional analysis, Rayleigh (1915) gave an example for the total, steady-state heat transfer rate $\dot{Q}$ from a long, thin cylinder of length L and radius a. He considered the wire and fluid as being characterized by single temperature each, respectively T and T_∞. See section 6.1.2 and the discussion of rule ND2 for additional details around this problem. For the assumption of an inviscid fluid flow, the following table shows the variables Rayleigh hypothesized as essential. The two rightmost columns provide the basic dimensions in Rayleigh's analysis as per this book and, in parenthesis, the basic dimensions considered by Rayleigh:

Variable	Definition	Physical quantity	
$\dot{Q}$	heat transfer rate (energy per time)	MLT^{-2}	(ET^{-1})
a	radius of wire	L	(L)
L	length of wire	L	(L)
$T - T_\infty$	cylinder-to-fluid temperature difference	Θ	(Θ)
ρc_p	specific heat of the fluid per volume	$ML^{-1}T^{-2}\Theta^{-1}$	$(EL^{-3}\Theta^{-1})$
V	velocity of fluid	LT^{-1}	(LT^{-1})
k	conductivity of fluid	$MLT^{-3}\Theta^{-1}$	$(EL^{-1}T^{-1}\Theta^{-1})$

Note that Rayleigh's dimensional analysis did not use the convention of L, T, M, and Θ that we have adopted. He did not use mass as a stand-alone dimension and used the following basic dimensions for the problem:

L, T, Θ, and E

where E is defined as the dimension of energy (here, thermal energy).

Rayleigh's approach exemplifies rule D1 in that, as per many steady-state heat transfer problems, Rayleigh expects the problem to scale not with two temperatures individually but as their difference, $T - T_\infty$. Second, Rayleigh expects the problem to scale not individually with either specific heat or density (of fluid) but only in terms of the product of these two, ρc_p. To continue with Rayleigh's analysis, his hypothesized function is then

$$\dot{Q} = f(a, \Delta T = T - T_\infty, V, \rho c_p, k).$$

Rule D1 applies as we have formulated in terms of known dependencies for temperature difference and the product of density and specific heat. Performing Ipsen's method (using either L, T, M, and Θ dimensions or with Rayleigh's L, T, Θ, and E), we can derive

$$\frac{\dot{Q}}{kL\Delta T} = F\left(\frac{aV\rho c_p}{k}, \frac{L}{a}\right).$$

We note that any such consistent change in the list of basic dimensions should yield a similar result. Next, Rayleigh reasons that, in the limit of large aspect ratio, the heat transfer will not depend on this the aspect ratio. As per rules ND3 and ND6, he proceeds to eliminate the aspect ratio and derive

$$\frac{\dot{Q}}{kL\Delta T} = G\left(\frac{aV\rho c_p}{k}\right).$$

Four months after Rayleigh's publication of March 18, 1915, Riabouchinsky (1911) wrote and published the following comment on Rayleigh's paper. Riabouchinsky's comment is aimed at the latter, simplified version of Rayleigh's formulation wherein Rayleigh considered only a single geometrical length a. In its entirety, Riabouchinsky's published comment reads as follows:

Lord Rayleigh gives this formula $\left[\frac{\dot{Q}}{ka\Delta T} = G\left(\frac{aV\rho c_p}{k}\right)\right]$ considering heat, temperature, length, and

time as four "independent" dimensions. If we suppose that only three of these quantities are really independent, we obtain a different result. For example, if the temperature is defined as the mean kinetic energy of the molecules, the principle of similarity allows us only to affirm that $\left[\frac{\dot{Q}}{ka\Delta T} = G\left(\frac{V}{ka^2}, \rho c_p a^3\right)\right]$.

Here, I quote Riabouchinsky's comment exactly except for the equations where we have used the current notation for clarity. He basically points out that, using kinetic theory, we can express temperature in terms of the kinetic energy of molecules, apparently removing the dimension of temperature from the problem. However, the result is an additional non-dimensional parameter! That is, additional physics seems to have resulted in a less powerful result and less insight. This issue has been termed the Rayleigh–Riabouchinsky paradox.

Indeed, this paradox seemed to puzzle even the great Rayleigh. Two weeks later, he responded to Riabouchinsky's comment by writing "it would indeed be a paradox if the further knowledge of the nature of heat afforded by molecular theory put us in a worse position than before in dealing with a particular problem.... It would be well worthy of discussion."

And so we here discuss it further.

11.2   Resolution of the Apparent Paradox

The so-called Rayleigh–Riabouchinsky paradox (where additional physics and removal of a variable increases the number of nondimensional parameters) has been discussed by several authors, notably Sedov (1950), Macagno (1971), and Cooper and West (1988). To resolve this conundrum, let's first look at Riabouchinsky's comment a little more closely. He proposed removing temperature by introducing the average kinetic energy of molecules. Namely, he proposes removing the dimension Θ using a definition of ΔT such that it is directly proportional to, and has dimension of, energy E.

Interestingly, as Sedov (1950) pointed out, Riabouchinsky has left out a *dimensional constant* in introducing kinetic theory, namely the Boltzmann constant, k_B. As per our discussion in section 5.1.3 and rule D3, this is inappropriate in dimensional analysis. The correct relation in kinetic theory between molecular kinetic energy and temperature has the following form:

Energy $\sim k_B T$, where $k_B = 1.38064852 \times 10^{-23}$ (Boltzmann constant).

Hence, the relationship between energy and T is not $E \sim \Delta T$ but $E/k_B \sim \Delta T$. Before continuing, let us here explore the implications of this. If we include the dimensional fundamental Boltzmann constant, then Riabouchinsky's list of variables would look something like this:

Variable	Definition	Physical quantity	
$\dot{Q}$	heat transfer rate (energy per time)	MLT^{-2}	(ET^{-1})
a	radius of wire	L	(L)
L	length of wire	L	(L)
$k_B(T - T_\infty)$	cylinder-to-fluid temperature difference	MLT^{-1}	(E)
$\rho c_p / k_B$	specific heat of the fluid per volume	L^{-3}	(L^{-3})
V	velocity of fluid	LT^{-1}	(LT^{-1})
k	conductivity of fluid	$L^{-1}T^{-1}$	$(L^{-1}T^{-1})$

Here again, the Rayleigh–Riabouchinsky dimensions are shown in parentheses. Hence, performing dimensional analysis, we have

$$\frac{\dot{Q}}{ka\Delta T} = G\left(\frac{k_B V}{ka^2}, \frac{\rho c_p a^3}{k_B}\right).$$

Note that I here used rule D1 in assuming the various groupings involving k_B and the product of density and specific heat. The expression is now dimensionally correct. However, this alone does not resolve the issue. The paradox's resolution is that Riabouchinsky's choice of variables has introduced irrelevant physics into the problem. Introduction of the microscopic motion of molecules into the problem is irrelevant. Indeed, the problem has been formulated in terms of continuum principles wherein time and space averages lead to the very concept of continuum quantities such as conductivity, density, specific heat, and temperature. Cherry-picking one continuum quantity such as temperature and then attempting to remove this continuum variable by "decomposing" it into molecular velocities is neither relevant nor useful. It introduces new physics that are not required.

Stated another way: The continuum velocity of the fluid is comparable to the time scales of advection and conduction of heat, the essence of the problem. Despite Riabouchinsky's use of the same variable, V, this velocity is very different (in magnitude and interpretation) from the mean average kinetic velocity of molecules. It also does not scale in the same way. For example, the conduction of heat scales as a temperature difference ΔT, while the kinetic molecular velocity is a function of the absolute temperature T. Introduction of such disparate physics takes away from the analysis. See notably Sedov (1950), Macagno (1971), and Cooper and West (1988) for discussions of this paradox.

11.3   The Rule of Relevance

I here attempt to generalize the lesson of the Rayleigh–Riabouchinsky paradox as follows:

> Dimensional analyses should include only relevant physics and should be careful to exclude irrelevant physics.

Relevance is, in turn, determined by the questions we ask. For Rayleigh's treatment of the Bousinessq problem, kinetic theory was irrelevant because Rayleigh asked a question about the heat rate (a volume-averaged concept) in terms of continuum temperature difference and as a function of continuum thermophysical quantities and macroscopic dimensions. This is much different from, say, asking about the average increase in the kinetic energy of individual gas molecules that touch the cylinder (for which we might invoke kinetic theory, Boltzmann constant, and the like).

In this book, we have in fact already touched on the subject of relevant physics in dimensional analysis. In section 5.4, we first asked a simple question regarding the time for a body near the surface of the Earth to fall from an initial height y_0, neglecting air resistance. To simplify the question and focus on relevant physics, we considered a uniform and constant value of acceleration of Earth's gravity (at its surface) g. We hypothesized a function of the form $t_0 = f_1(h, g, m)$, and the result of dimensional analysis was simply

$$gt_0^2/h = c_0.$$

This was a succinct summary of the relevant physics, and an exclusion of *irrelevant physics*.

In sections 5.5 and 5.6, we then changed the question and broadened the scope of what is *relevant physics*. Namely, we asked about the time scales of orbital periods of interacting celestial bodies (of immense mass) across vast distances. Here we kept G as per rule D3 and considered

$$t_0 = f_1(h, r, G, m, M).$$

Dimensional analysis then led us to

$$\frac{GMt_0^2}{r^3} = F\left(\frac{h}{r}, \frac{m}{M}\right).$$

From this, we eventually gleaned some version of Kepler's third law.

The key to the latter example is that the new question prompted newly relevant physics of celestial motion, and this led to a new treatment and new variables. It prompted additional complexity.

Bridgman (1911) also discusses the resolution of the Rayleigh–Riabouchinsky paradox as follows. Bridgman's argument is basically that all problems of dimensional analysis involve selection of a sufficient set of physical quantities to describe that problem. The list is then ultimately validated by how accurately the list models the physical situation. In deciding which quantities to neglect, Bridgman writes, "the justification will involve real argument and a considerable physical experience with physical systems of the kind which we have been considering. The problem cannot be solved by the philosopher in his armchair, but the knowledge involved was gathered only by someone at some time soiling his hands with direct contact." In other words, the "proof" that any list of variables is correct and sufficient is ultimately how well (and how sufficiently) the resulting theory agrees with experimental observations.

11.4 Summary of Rules of Thumb for Combining Physical Insight with Dimensional Analysis

In sections 5.1, 6.1, and 11.2, I proposed a set of rules of thumb useful in dimensional analysis. Table 11.1 summarizes these rules:

Table 11.1
Rules of thumb for dimensional analysis proposed in this book

No., Relevance	Rule	Example
D1. Dimensional variables	If the variable consistently appears as some grouping (e.g., algebraic), then use the group as an independent variable.	For spring and combined mass problem, hypothesize function as $(x - x_0) = f(m_h + m_b, k, g)$
D2. Dimensional variables	Exclude a variable if you can express it as a function of the others.	Write $r = f(x, y, z)$ as $r = f(x, y)$ if $z = g(x, y)$.
D3. Dimensional variables	Keep dimensional constants, but absorb nondimensional constants.	Einstein: $E = f(m, c)$, not $E = f(m)$ Circle: $A = f(r)$, not $A = f(\pi, r)$
D4. Dimensional variables	Exclude any variable that involves a unique dimension.	Given $s = f(x, y, z)$, if only z contains dimension of mass, then $s = f(x, y)$.
D5. Dimensional variables	If invoked, leverage geometric similarity.	Take into account any length s, area A, and volume V simply with $f(s)$.
ND1. Nondimensional Variables	Reorganize expressions of nondimensional variables (but maintain the same number of parameters).	We can write $H = F_1(X, Y, Z)$ as $H^a = F_2(XY^b, Y, Z)$ for nonzero a, b. Also $1 = F(X)$ yields $X =$ constant.
ND2. Nondimensional Variables	Isolate, then evaluate, a known dependence (if other variables are not affected).	Given $r/d = G(a/b, a^2/c^2)$, and $r \propto b^s$, we might conclude $R = \left(\frac{b^p}{a^s}\right) H\left(\frac{a^2}{c^2}, \frac{d}{e}\right)$.
ND3. Nondimensional Variables	Isolate a variable with an unknown but weak dependence.	For a weak dependence on V/c: Never $F/\rho c^2 b^2 = F_1(\rho cb/\mu, V/c)$; instead $F/\rho V^2 b^2 = F_1(\rho Vb/\mu, V/c)$.
ND4. Nondimensional Variables	Nested function of nondimensional parameters.	Simplify $A = F(B, C, D)$ to $A = F(C, D)$ if $B = G(C, D)$
ND5. Nondimensional Variables	Consider absorbing into function a nondimensional parameter on which the dependent variable has a very weak relative dependence.	For $A = F(B, C, D)$ write $A \cong F(B, C)$ if dependence on D is known to be very weak.
ND6. Nondimensional Variables	Be very careful eliminating variables—even if they are small.	$A = F(B, C, D)$ for small D could be of some form $A \approx CD + BD$, hence we cannot eliminate D.
Rule of relevance	Include relevant physics, exclude irrelevant.	$t_0 = f_1(h, g, m)$ for a rock on Earth, but $t_0 = f_1(h, r, G, m, M)$ for Earth vs. the sun.

12

Common Dimensionless Groups

A colour is a physical object as soon as we consider its dependence, for instance, upon its luminous source, upon other colours, upon temperatures, upon spaces, and so forth.
—Ernst Mach (1914)

In addition to the numbers C_D, Re, We, and Fr, which we have discussed in most detail, we further summarize some common dimensionless groups in the next two sections and in table 12.1.

12.1 Mach Number

The Mach number applies to high-speed (compressible) flows and can be formulated as follows:

$$\mathrm{Ma} = \frac{V}{c}.$$

V = local fluid velocity

c = local speed of sound.

Squaring this ratio and introducing density and length, we have a ratio of a kinetic energy per unit fluid volume (i.e., a dynamic pressure) to the potential energy stored in the compressible fluid:

$$\mathrm{Ma}^2 = \frac{V^2}{c^2} = \frac{\rho}{\rho}\frac{V^2}{c^2}\frac{L^2}{L^2}.$$

Consider that c can be well approximated in terms of an isentropic change in pressure with respect to density as

$$c = \sqrt{\left.\frac{\partial p}{\partial \rho}\right|_{\Delta s=0}}.$$

Table 12.1
Examples of dimensionless groups (adapted from Yarin, 2012)

Name	Symbol	Formulation	Comparison ratio	Field of use
Biot	Bi	hL/k	Convection heat transfer to conduction heat transfer	Heat transfer
Bond	Bo	$\rho g L^2/\sigma$	Gravity force to capillary force	Motion of drops, bubbles, wicking
Brinkman	Br	$\mu V^2/(k\Delta T)$	Heat dissipation to heat transfer	Viscous flow, heat transfer
Capillary	Ca	$\mu V/\sigma$	Viscous force to capillary force	Multiphase flow
Damkohler	Da	kL^2/D	Reaction rate to transport rate (diffusive rate shown)	Reacting flows
Dean	De	$VL^{3/2}\rho/(\mu r^{1/2})$	Centrifugal and inertial force relative to viscous force	Flows in curved channels
Eckert	Ec	$V^2/(c_p\Delta T)$	Kinetic energy to thermal energy	High speed flows
Euler	Eu	$\rho V^2/(\Delta p)$	Dynamic pressure to pressure drop/difference	Aerodynamics, channel flow
Froude	Fr	$V/\sqrt{gL}$	Inertia to gravity force, kinetic energy to potential energy	Wave motion, biomechanics
Grashof	Gr	$\rho^2 g\beta L^3\Delta T/\mu^2$	Buoyancy force to viscous force	Natural convection
Knudsen	Kn	λ/L	Mean free path to characteristic dimension	Rarefied gas flow
Lewis	Le	$k/(\rho c_p D)$	Thermal diffusivity to molecular diffusivity	Combined heat/mass transport
Mach	Ma	V/c	Inertia to compressibility forces	High speed flows
Peclet	Pe	VL/D	Advection rate to diffusive rate	Convection, mass transport
Prandtl	Pr	$\mu c_p/k$	Momentum diffusivity to thermal diffusivity	Heat transfer in fluid flow
Reynolds	Re	$\rho LV/\mu$	Inertia force to viscous force, dynamic pressure to viscous pressure	Fluid mechanics
Schmidt	Sc	$\mu/(\rho D)$	Momentum diffusion to molecular diffusion	Mass transport
Weber	We	$\rho V^2 L/\sigma$	Dynamic pressure to capillary pressure	Bubble formation, drop impact

Hence,

$$\text{Ma}^2 = \frac{\rho}{\rho}\frac{V^2}{\left.\frac{\partial p}{\partial \rho}\right|_{\Delta s=0}}\frac{L^2}{L^2} = \frac{\text{kinetic energy per volume}}{\text{potential energy (per volume) stored in compressed fluid}}.$$

12.2   Euler Number

The Euler number (aka pressure coefficient, C_p) is defined as

$$Eu(= C_p) = \frac{p - p_\infty}{\frac{1}{2} \rho_\infty V_\infty^2} = \frac{\text{net pressure change}}{\text{dynamic pressure}}.$$

$\rho_\infty V_\infty^2$ = dynamic pressure of approaching flow

p_∞ = static pressure of approaching flow

p = local pressure

The Euler number is especially important for (nearly inviscid) flows at sufficiently high Reynolds numbers and in regions outside of boundary layers and wakes.

12.3   Table of Dimensionless Groups

Table 12.1 presents a list of common nondimensional groups and a summary of their interpretations and fields of use. The formulations in the table use common variables for dimensional scales including L, v, T, p, g, and λ for characteristic length, velocity, temperature, pressure, gravitational acceleration, and mean free path, respectively. Thermophysical properties used include respectively k, c_p, μ, β, and σ for conductivity, specific heat, dynamic viscosity, coefficient of thermal expansion, and surface tension. Also used are speed of sound, c, and heat transfer coefficient, h.

Depending on the problem, it can be very useful to classify nondimensional parameters in this table as ratios of one type of the following quantities: forces, energy, transport rates, lengths, or pressures. For example, our discussions around Reynolds number have mostly interpreted it as a ratio of inertial to viscous forces. However, in some cases (e.g., unsteady flow problems), it may be useful to think of Reynolds number as a ratio of the advective transport of momentum versus the diffusion of momentum via viscosity. Consider that the two directions of this transport can be orthogonal (so the force ratio comparison is less useful). Weber number is most clearly a ratio of two types of pressure or pressure forces. Knudsen number is most clearly interpreted as a length ratio. Among Schmidt, Prandtl, and Lewis numbers, there are two independent nondimensional parameters, and all involve groupings of thermophysical properties. These groupings are perhaps best described as characterizing transport rates among transport of momentum (via viscosity), molecular diffusion (via Brownian motion of molecules), and storage and diffusion (via phonon transport) of thermal heat.

12.4   Summary

- We presented physically intuitive descriptions of the nondimensional Mach and Euler numbers.

- We presented a table of common nondimensional parameters including their common formulation and physical interpretation.

- Physical interpretations of common dimensional groups include ratios of forces, energy, transport rates, and pressures.

Problems

12.1. Show that another version of Peclet number can be derived by combining the Peclet number of table 12.1 with the Lewis number shown in the table. This Peclet number is applicable to convective heat transfer phenomena.

12.2. We can think of a generalization of the Peclet number as describing the ratio of advection of momentum, heat, or species to the diffusion of these respective quantities. Show that the Reynolds number can therefore be interpreted as a form of Peclet number. Consider a product of Schmidt and Reynolds numbers and discuss briefly.

12.3. Dean number De can be used to characterize fluid flows in circular pipes of diameter d whose centerline is curving through a circular arc of radius r. Show that Dean number can be expressed as the geometric mean of inertial forces (in streamwise direction) and centripetal forces (perpendicular to the streamwise direction).

12.4. Derive a more interesting version of capillary number where surface tension forces and viscous forces are each characterized by different length scales x and y, respectively.

13

Scaling Using Approximate Equations

…eine reine Zahl, die wir die *Reynolds'sche Zahl* nennen wollen. [R is a pure number; we will call it the Reynolds number.]

—Arnold Sommerfeld (1908)

We have looked at the application of dimensional analysis to problems where we have enough intuition and/or understanding to identify essential variables, but we have not (for the most part) striven to write the governing equations or to scale such equations. In some cases (e.g., running dinosaurs, flying birds, the biomechanics of weightlifters and rowers), the governing equations are either unknown or extremely difficult to solve. In other cases, we have ignored our knowledge of the basic equations in order to test and demonstrate the power of dimensional analysis. But in the latter case, we only feigned ignorance to learn about dimensional analysis.

Observing and developing an understanding of the physical world to the point where we can discover and test (by experimental observation) governing principles and laws that predict behavior is an essential (perhaps the essential) goal of science. Even a basic understanding of governing equations is often superior to simple dimensional analysis alone. By "basic understanding," consider that you may be able to draft a governing equation for a process that you cannot resolve analytically. The equations may be algebraic relations or may be ordinary or partial differential equations (PDEs).

Deriving nondimensional parameters from governing equations is always superior to deriving from dimensional parameters alone. The number of nondimensional parameters resulting from scaling of governing equations is always equal to or less than those obtained from analyzing the variables alone.

In this chapter, we discuss the scaling of governing equations by revisiting the example of the Olympic rowboats (sculls) from chapter 7. In section 7.4, we applied our step-by-step approach of dimensional analysis to McMahon's (1971) problem, and then applied the rules summarized in table 11.1 to introduce physical insight. We note that McMahon himself did not approach the problem in this way—instead, he hypothesized a set of proportionality relations that respected the dependence of variables but not the principle of dimensional homogeneity (PDH). For example, McMahon expresses velocity dependence on boat length as $V^3 \propto n/L^2$, and so on. He then combines these proportionalities to derive his final result of $V \propto n^{1/9}$.

Here, I will take a different tack. I identify algebraic relations among key variables but will strive to work with relations that respect PDH. If an exact relation is not known, then I will define and work with appropriate (and verifiable) constants for these relations. We will also more explicitly consider geometric similarity.

We first invoke Archimedes' principle to relate total weight to the geometry of the displaced volume (of length s). From Archimedes' principle, the weight of the displaced water can be expressed as

$$\text{displaced water weight} = c_0 \rho s^3 g = (nW_{\text{boat}} + nW_0) = n(W_{\text{boat}} + W_0).$$

Here, the dimensionless constant c_0 takes into account the shape of the particular geometry of the submerged portion of the boat (geometric similarity of the submerged geometries demands all scales have same value of c_0). Rearranging,

$$s^3 = \frac{(W_{\text{boat}} + W_0)}{c_0 \rho g} n \Rightarrow s^2 = \left(\frac{(W_{\text{boat}} + W_0)}{c_0 \rho g} n \right)^{2/3}.$$

We now leverage our knowledge of drag laws as discussed throughout this book, particularly in chapters 4, 7, and 10. The drag laws for geometrically similar bodies suggest that the total rowing power of the boat nP_0 in terms of a boat drag F_D can be expressed in terms of a drag coefficient C_D as

$$nP_0 = F_D V = C_D \frac{\rho V^3 c_1 s^2}{2}.$$

The dimensionless constant c_1 here accounts for whichever particular area best characterizes the boat drag. For example, McMahon insightfully suggests the hull's wetted area, consistent with our discussion section 10.3. Here, we consider the geometric similarity of the immersed geometry and, naturally, take into account the wetted area of the boat. Note that, for this rowing analysis, the specific choice of area is unimportant so long as there is at least one that makes C_D approximately constant (e.g., insensitive to Reynolds number). Combining our Archimedes and drag relations to eliminate length scale:

$$nP_0 = C_D \frac{\rho V^3 c_1}{2} \left(\frac{(W_{\text{boat}} + W_0)}{c_0 \rho g} n \right)^{2/3}.$$

Solving for V in terms of n (our original question), we arrive at

$$C_D \, \frac{\rho V^3 c_1}{2P_0} \left(\frac{(W_{\text{boat}} + W_0)}{c_0 \rho g} \right)^{2/3} = \frac{n}{n^{2/3}} = n^{1/3}.$$

The latter equation is nondimensional. We can algebraically rearrange this as

$$V = \left(\frac{2P_0}{\rho C_D c_1} \left(\frac{(W_{\text{boat}} + W_0)}{c_0 \rho g} \right)^{-2/3} \right)^{1/3} n^{1/9}.$$

We see the same $V \propto n^{1/9}$ result as before, but in the current analysis we have kept additional information regarding the exponents of weight, the drag law, and so forth. Importantly, our approach (unlike McMahon's approach) has resulted in an equation that explicitly includes the effects of viscosity via the drag coefficient. Our more general result helps explore the trade-offs and dependencies among drag coefficient reduction, the rowing power contributed by any one rower, and weight reduction. The current result also clearly identifies a natural rower quality measure of the form $P_0^{1/3}(W_{\text{boat}} + W_0)^{-2/9}$, which may be useful in analyzing the trade-off between output power and weight of potential rowers. Dimensional analysis and scaling of even approximate equations governing the physics is indeed a powerful tool with great potential for insight.

13.1   Summary

- Throughout this book, we have performed dimensional analysis by identifying essential variables in a problem and then analyzing the relation among them. We have also striven to identify rules and guidelines by which physical intuition can be brought to bear. In this chapter, we introduced a separate and powerful method based on analysis of approximate governing equations.

- Analysis of even approximate governing equations can be superior to more standard dimensional analysis alone because the form of the equations already limits and specifies relations among variables.

- We demonstrated this approach by revisiting the rowing example from chapter 7 to derive a scaling law. The analysis is more intuitive than that presented by McMahon in that it preserves PDH (in contrast to simple proportionality statements, which do not). In addition, our analysis was arguably more complete than McMahon's in that it resulted in a scaling law that considers the effect of viscosity (through the drag coefficient) and includes verifiable proportionality constants in the final relation. Our relation also identifies trade-offs among drag coefficient reduction, weight reduction, and power output per weight for the rowers.

- Deriving nondimensional parameters from governing equations is typically superior to deriving from dimensional parameters alone. The number of nondimensional parameters resulting from scaling of governing equations is always less than or equal to those obtained from analyzing the variables alone.

Problems

13.1   The equation of motion for falling rock and neglecting air resistance is given by

$$y = y_0 + v_0 t + \frac{1}{2} g t^2.$$

Nondimensionalize this equation using the respective initial height and velocity, y_0 and v_0, as scaling constants. Identify any dimensionless parameters that arise.

13.2.   Conservation of energy states that the summation of kinetic energy and potential energy is constant (e.g., neglecting the heat generated by friction). We express this for some mass as

$$\frac{1}{2} m v^2 + mgh = \text{constant}.$$

Nondimensionalize the equation using the respective characteristic velocity and height, v_0 and h_0, as scaling constants. Identify any dimensionless parameters that arise.

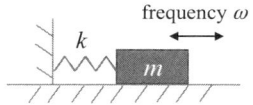

13.3.   Consider the following equation for the time t response $x(t)$ of an underdamped spring-mass system with a damping coefficient γ (due to friction), a natural frequency ω, and a phase α, which can be written as

$$x(t) = x_0 e^{-\gamma t} \cos(\omega t - \alpha),$$

where x_0 is the initial magnitude of the displacement. Nondimensionalize this equation and consider the limiting behavior as the argument of the exponent (which must itself be nondimensional) becomes small but not zero.

Closing Note

The analyses and various examples presented in this book should serve as an introduction to the general approach and value of dimensional analysis and provide some guidance as to how to combine dimensional analysis with physical insight. My hope is that readers will try their luck at posing a new scaling law and validating it with experiments.

Appendix A: Properties of Common Fluids

Table A.1
Fluid properties at 1 atm and 20°C (adapted from White, 2016)

Fluid	μ, kg/(m·s)*	Ratio, μ/μ (H$_2$)	ρ, kg/m³	ν, m²/s†	Ratio, ν/ν (Hg)
Hydrogen	8.8 E−6	1.0	0.084	1.05 E−4	920
Air	1.8 E−5	2.1	1.20	1.51 E−5	130
Gasoline	2.9 E−4	33	680	4.22 E−7	3.7
Water	1.0 E−3	114	998	1.01 E−6	8.7
Ethyl alcohol	1.2 E−3	135	789	1.52 E−6	13
Mercury	1.5 E−3	170	13,580	1.16 E−7	1.0
SAE 30 oil	0.29	33,000	891	3.25 E−4	2,850
Glycerin	1.5	170,000	1,264	1.18 E−3	10,300
Sea Water	~1.08E−3		~1027		

*1 kg/(m·s)=0.0209 slug/(ft·s)

†1 m²/s = 10.76 ft²/s

Temperature dependence of water viscosity (from Fox et al., 2015):

$$\mu_{water} \cong (2.414E-5)10^{[247.8\,\mathrm{K}/(T-140\,\mathrm{K})]} \pm 2.5\% \text{ over } 0°C \text{ to } 370°C$$

References

Achenbach, E. The effects of surface roughness and tunnel blockage on the flow past spheres. *Journal of Fluid Mechanics* 65.1 (1974): 113–125.

Achenbach, E. Experiments on the flow past spheres at very high Reynolds numbers. *Journal of Fluid Mechanics* 54.3 (1972): 565–575.

Adam, C. D. Fundamental studies of bloodstain formation and characteristics. *Forensic Science International* 219.1–3 (2012): 76–87.

Alexander, M. R. A dynamic similarity hypothesis for the gaits of quadrupedal mammals. *Journal of Zoology, London* 201 (1983): 135–152.

Alexander, M. R. Walking and running. *The Mathematical Gazette* 80 (1999): 488.

Alexander, R. M. *Dynamics of dinosaurs and other extinct giants*. New York: Columbia University Press, 1991.

Almedeij, J. Drag coefficient of flow around a sphere: Matching asymptotically the wide trend. *Powder Technology* 186.3 (2008): 218–223.

American Society of Mechanical Engineering. Historic mechanical engineering landmark: The David Taylor Model Basin. Naval Surface Warfare Center, Carderock Division, West Bethesda, Maryland, January 30, 1998.

Attinger, D., et al. Fluid dynamics topics in bloodstain pattern analysis: Comparative review and research opportunities. *Forensic Science International* 231.1 (2013): 375–396.

Batchelor, G. K. *The life and legacy of G. I. Taylor*. Cambridge: Cambridge University Press, 2008.

Belendez, A., C. Pascual, D. I. Mendez, T. Belendez, C. Neipp. Exact solution for the nonlinear pendulum. *Revista Brasileira de Ensino de Fisica* 29.4 (2007): 645–648.

Berberan-Santos, M. N., and L. Pogliani. Two alternative derivations of Bridgman's theorem. *Journal of Mathematical Chemistry* 26.1–3 (1999): 255.

Blevins, R. D. *Applied fluid dynamics handbook*. New York: Van Nostrand Reinhold, 1984.

Boussinesq, J. V. Comptes rendus. *Journal de Mathematiques* 1 (1905): 285–332.

Bridgman, P. W. *Dimensional analysis*. New Haven, CT: Yale University Press, 1931.

Chandkrasekhar, S. *Newton's Principia for the common reader*. Oxford: Oxford University Press, 1995.

Choi, J., W.-P. Jeon, and H. Choi. Mechanism of drag reduction by dimples on a sphere. *Physics of Fluids* 18.4 (2006): 041702.

Coleman, H. W., and W. G. Steele. *Experimentation and uncertainty analysis for engineers*. Hoboken, NJ: John Wiley, 1998.

Colino, J. M., A. J. Barbero, and F. J. Tapiador. Dynamics of a skydiver's epic free fall. *Physics Today* (April 2014): 64–65.

Cooper, N. G., and G. B. West, eds. *Particle physics: A Los Alamos primer*. Cambridge: CUP Archive, 1988.

Deakin, M. A. B. G. I. Taylor and the Trinity test. *International Journal of Mathematical Education in Science and Technology* 42 (2011): 1069–1079.

Duan, Z., B. He, and Y. Duan. Sphere drag and heat transfer. *Scientific Reports* 5 (2015): 12304.

Fey, U., et al. A new Strouhal-Reynolds-number relationship for the circular cylinder. *Physics of Fluids* 10.7 (1998): 1547.

Fourier, J. *Analytical theory of heat.* Translated by Firmin Didot. Cambridge University Press, 1822.

Fox, R. W., A. T. McDonald, and P. J. Pritchard. *Introduction to fluid mechanics.* Ninth edition. Hoboken, NJ: John Wiley & Sons, 2015.

Fulcher, L. P., and B. F. Davis. Theoretical and experimental study of the motion of the simple pendulum. *American Journal of Physics* 44.1 (1976): 51–55.

Gibbings, J. C. *Dimensional analysis.* London: Springer, 2011.

Hirt, C., S. Claessens, T. Fecher, M. Kuhn, R. Pail, and M. Rexer. New ultrahigh-resolution picture of Earth's gravity field. *Geophysical Research Letters* 40.16 (2013): 4279–4283.

Incropera, F. P., A. S. Lavine, T. L. Bergman, and D. P. DeWitt. *Principles of heat and mass transfer.* Seventh edition. Hoboken, NJ: John Wiley, 2013.

Ipsen, D. C. *Units, dimensions, and dimensionless numbers.* New York: McGraw-Hill, 1960.

Lemon, D. S. *A student's guide to dimensional analysis.* Cambridge: Cambridge University Press, 2017.

Lima, F. M. S., and P. Arun. An accurate formula for the period of a simple pendulum oscillating beyond the small angle regime. *American Journal of Physics* 74.10 (2006): 892–895.

Macagno, E. O. Historico-critical review of dimensional analysis. *Journal of the Franklin Institute* 292.6 (1971): 391–402.

Mach, E. *The analysis of sensations, and the relation of the physical to the psychical.* Chicago: Open Court, 1914.

Mack, J. E. Semi-popular motion-picture record of the Trinity explosion. U.S. Atomic Energy Commission, MDDC-221, 1946.

Mahajan, S. *Street-fighting mathematics.* MIT OpenCourseWare, 2008.

Maxwell, J. C., and J. J. Thompson. *A treatise on electricity and magnetism*, vol. 1. Clarendon, 1892.

McMahon, T. A. Rowing: A similarity analysis. *Science* 23 (July 1971): 349–351.

Mills, I., T. Cvitaš, K. Homann, N. Kallay, and K. Kuchitsu. *Quantities, units and symbols in physical chemistry.* Hoboken, NJ: Blackwell Science, 1993.

Moulton, F. R. The influence of astronomy on mathematics. *Science* (1911): 357–364.

Newton, I. *Newton's Principia: The mathematical principles of natural philosophy.* Trans. A. Motte. New York, NY: Geo. P. Putnam, 1850.

Nikuradse, J. Gesetzmäßigkeiten der turbulenten Strömung in glatten Rohren (Nachtrag). *Forschung im Ingenieurwesen* 4.1 (1933): 44.

Nikuradse, J. Laws of flow in rough pipes. National Advisory Committee for Aeronautics, Technical Memorandum 1292, 1950.

Rainforth, E. C., and M. Manzella. Estimating speeds of dinosaurs from trackways: A re-evaluation of assumptions. Contributions to the paleontology of New Jersey (II)–Field guide and proceedings, 2007 Oct. 12–13; Geological Association of New Jersey XXIV Annual Conference and Field Trip, 2007.

Rayleigh, L. The principle of similitude. *Nature* 95 (1915): 66.

Riabouchinsky, D. The principle of similitude: Letter to the editor. *Nature* 95 (1915): 591.

Roth, E. Drag of non-spherical solid particles of regular and irregular shape. *Powder Technology* 182 (2008): 342–353.

Rott, N. Lord Rayleigh and hydrodynamic similarity. *Physics of Fluids* A 4.12 (December 1992), 2595–2600.

Russell, J. L. Kepler's laws of planetary motion: 1609–1666. *British Journal for the History of Science* 2.1 (1964): 1–24.

Schlichting, H., and K. Gersten. *Boundary-layer theory.* Berlin: Springer, 2000.

Sedov, L. I. 1950. *Similarity and dimensional methods in mechanics.* New York: Academic Press, 1959.

Sommerfeld, A. Ein Beitrag zur hydrodynamischen Erklärung der turbulenten Flüssigkeitsbewegung. *Proceedings of the International Congress on Mathematicians (Rome, April 6–11, 1908)* 3 (1909):116–124.

Taylor, G. I. The formation of a blast wave by a very intense explosion. I. Theoretical discussion. *Proceedings of the Royal Society of London A* 201.1065 (1950a): 159–174.

Taylor, G. I. The formation of a blast wave by a very intense explosion. II. The atomic explosion of 1945. *Proceedings of the Royal Society of London A* 201.1065 (1950b): 175–186.

Taylor, M., A. I. Diaz, L. A. Jodar-Sanchez, and R. J. Villanueva-Mico. 100 Years of dimensional analysis: New steps toward empirical law deduction. arXiv preprint. arXiv:0709.3584.

Tennekes, H. *The simple science of flight.* Cambridge, MA: MIT Press, 1997.

Van Dyke, M. *An album of fluid motion.* Stanford, CA: Parabolic Press, 1982.

White, F. M. *Fluid mechanics.* Eighth edition. New York: McGraw-Hill, 2016.

Wright, O., and F. C. Kelly. *How we invented the airplane.* Philadelphia: McKay, 1953.

Yarin, L. P. *The pi-theorem.* Berlin: Springer, 2012.

Zohuri, B. *Dimensional analysis and self-similarity methods for engineers and scientists.* Berlin: Springer, 2015.

Index

Aerodynamics
 airfoils, 32–33, 135–136
 bluff bodies, 135–136
 falling body, 35–38
 of flyers, 100–107. *See also* Flight; Effect of
  streamlining
 lift and drag coefficients, 101
Ambiguities arising from incomplete dimensional
  analysis, 137–138
Archimedes' principle, 94, 152
Astrophysics
 orbital period of planets, 57
 universal gravitational constant, 44, 55–57.
  See also Kepler's third law; Orbital period
Atomic explosion analysis, 78–81
 Trinity bomb test, 79–80

Baumgartner, Felix, 38
Bench press. *See* Weightlifters
Bingham number, 129
Biomechanics
 muscle stress, 82
 of flight, 99–105
 of rowers, 97–98
 of weightlifters, 82–84
 walking and running, 84–86
Blood splatter (forensics), 74
Boats
 drag on row boat (racing shell), 93–99, 151–153
 drag on ships, 125–128
Bodies falling through air
 bluff body, 135
 skydiver, 35
 stone (or ball), 50–54
Boltzmann's constant, 143
Bond number (*Bo*), 91
Boundary layers
 laminar versus turbulent, 22–23, 131–132
 separation, 131–134
Brute force experimentation, 29–31
Buckingham, Edgar, 2

Buckingham pi theorem
 comparison to Ipsen's method, 119
 method, 111–115
Closing a problem with experiments, 50–54,
  61–62, 116–118
Commercial jet, 101–105
Collapse of data
 atomic explosion radius, 80
 droplet splatter spine data, 76
 falling body, 53
 general concept, 34–35
 mechanical and animal flyers, 104
 orbital periods, 57
 pendulum periods, 62
 rower velocity data, 99
 running animals, 88
 weightlifter data, 84. *See also* Scaling
Conservation of energy, 20

Darcy friction factor, 116
Dead lift. *See* Weightlifters
Determinant, 113
Dimensional analysis. *See* Ipsen's method
Dimensional functions, rules for. *See* Rules for
  combining physical insight with dimensional
  analysis
Dimensional matrix, 113–114
Dimensionless numbers, 148
Dimensions, 6–9
 of derivatives, 8
 of integrals, 8
 versus units, 6
Dinosaur speed, 84–89
 table of data, 89
 Tyrannosaurus rex (*T. rex*), 89
Drag
 on a cylinder, 32–33
 drag crisis, 134
 effect of streamlining, 32–33
 on flyers, 105–108

Drag (cont.)
  power dissipated, 29, 97, 105–106
  on rough spheres, 131–135
  on smooth spheres, 33–35, 133–135
  on a submarine, 29–31, 47–49
  surface roughness effect on, 45, 131–135.
    See also No-slip condition
Drag coefficient (C_D), 32–37, 48–49, 101–102,
  105, 135–136
Drag crisis, 134
Dynamic pressure, 22–23
Dynamic similarity, 121–129

Einstein, Albert, 43
Effect of streamlining, 32–33
Euler, Leonhard, 1
Euler number (pressure coefficient, *Eu*),
  149–150

Fission bomb. *See* Atomic explosion analysis
Flight
  aircraft engine power, 103
  angle of attack, 32
  and drag coefficient, 48–49, 101, 105
  great flight diagram, 99, 102–105
  human powered, 103
  insects and mosquitos, 105
  lift coefficient (C_L), 107
  wing span and area, 28, 100–103
Fluid mechanics, 21–24. *See also* Aerodynamics;
  Boats; Drag; Flight; Pipe flow
Fourier, Jean-Baptiste, 6
Friction factor. *See* Darcy friction factor
Froude number (*Fr*), 86–89, 96
  in biomechamics, 86–89
  in wave motion, 96
Functions, 9
  asymptotic limits of, 14–15
  function of a constant, 10
  inversion of, 10–11, 87, 108
  nested function, 11
  range of applicability, 11
  unknown functions, 9–10. *See also* Rules for
  combining physical insight with dimensional
  analysis

Gait, 85–87
  running vs. walking, 85–86
Geometric similarity, 28–29, 33
  relevance to absorbing dimensionless constants,
  44–45
Guessing unknown functions, 13–14

Heat transfer
  between wire and flow, 70–71, 141–143
  continuum versus noncontinuum, 144
Hydraulics. *See* Fluid mechanics; Pipe flow

Ipsen, David Carl, 3, 45
Ipsen's method
  comparison with Buckingham pi theorem, 119
  description and comments, 45–47
  detailed examples of, 48, 52, 56, 95, 100–101
  inelegant applications of, 49–50

Kepler's third law, 57
Kinematic similarity, 121–122
Kinetic energy, 20
Kinetic theory, 143
Knudsen number (*Kn*), 148–149

Laminar flow, 23–24
Lewis number (*Le*), 148–149
Liquid droplets
  splatter and spine patterns, 74–77. *See also* Weber
  number (*We*)
Lift coefficient (C_L), 32, 101–102, 107

Mach number (*Ma*), 31, 38, 71–72, 105,
  147–149
Martini glass, 13–14
Maxwell, James Clerk, 6, 112
McMahon, Thomas A., 93
Midge, 100–101, 104
Models and prototypes, 28–29, 121–128
Momentum, 19–20
Muscle stress, 82

National Advisory Committee for Aeronautics
  (NACA), 32
Newton, Isaac, 1, 2, 5, 19, 20, 21, 22
Newtonian fluid, 21
Newton's second law, 1, 19, 20, 22, 24, 25,
  36, 51
No-slip condition, 21–22
Nondimensional functions, rules for. *See* Rules for
  combining physical insight with dimensional
  analysis

Olympic athletes. *See* Weightlifters; Rowers
Orbital period, 56–57, 145

Peclet number (*Pe*), 148, 150
Pendulum, 59–63
  measuring Earth sphericity, 63
  upside-down, 85
Pi (π), 44
Pipe flow, 112–113, 116–118
  Moody chart for, 117
  proportionality to pipe length, 116
  roughness effect on, 112–113, 117–118
Potential energy, 20–21
Power
  of flyers, 105–106
  of rowers, 152–153

Power law
  atomic explosion radii, 80
  droplet splatter spine data, 76
  mechanical and animal flyers, 104
  orbital periods, 57
  as possible asymptotic limit, 15
  rower velocity data, 99
Prandtl number (Pr), 148–149
Principle of dimensional homogeneity (PDH), 12
  in approximate equations, 152
  in evaluating a known dependence, 70
  guessing functions, 13
  spotting errors, 12–13

Rayleigh, Lord, 1–2, 70–71, 131, 141–143
Reynolds number (Re), 23–25
  critical (transition), 24
  as dependent variable in boat drag, 96
  determining size of wake, 132–133
  drag dependence on, 35, 134
  in droplet splatter, 77
  in flight, 101
  in flight power dissipation, 105–106
  forcing formulation to include, 72
  interpretations of, 32, 48–49, 149
  inverse scaling to Froude number (Fr), 127–128
  laminar versus turbulent, 23–24, 132, 134, 117
  lift dependence on, 106–107
  in prototype scaling, 124–125
  versus roughness for bluff body drag, 133–135
Riabouchinsky, Dimitri, 2, 141–142,
Riabouchinsky–Rayleigh Paradox, 141–143
  resolution of, 143–144. See also Rule of relevance
Rowers, 93–99, 151–153
Rule of relevance, 144–146
Rules for combining physical insight with
    dimensional analysis
  for functions of dimensional variables (rules
    D1–D5), 41–45
  for functions of dimensionless variables (rules
    ND1–ND6), 67–73
  rule of relevance, 144–145
  summary of, 146
Running speed of animals, 85–90

Scaling, 121–128
  confirmation of scaling of atomic explosion data,
    78–81
  difficulty in confirming a weak scaling, 98–99.
    See also Collapse of data
Schmidt number (Sc), 148–149
Sedov, Leonid, 78
Similarity laws, 122–123
Sommerfeld, Arnold, 151
Speed of light, 43–44
Speed of sound, 71, 72, 91, 105, 147, 148.
    See also Mach number (Ma)

Spine patterns (droplets). See Liquid
    droplets
Splatter, 75–76
Sports science
  Olympic weightlifters, 82–84
  rowers, 93–98, 151–153
Strouhal number (St), 65
Submarines, 29–32
  Ipsen's method applied to, 47–49

Taylor, Geoffrey Ingram, 78
Tennekes, Hendrik, 9, 93, 104–105
Terminal velocity, 37–38
Thermophysical properties, 148–149, 157
Turbulence, 23–24. See also Turbulent
    flow
Turbulent flow, 23–24
  boundary layer, 22–23, 131–134
  laminar-to-turbulent transition, 23–24
  pipe flow, 116–119
  wake, 131–135

Unit conversion factors (UCFs), 8–9
Units, 6–7
  Système internationale d'unités, 6
Universal gravitational constant, 55

Vaschy, Aimé, 2
Von Neumann, John, 78

Weber number (We), 75–77
Weightlifters, 82–84, 109